# Net Centricity and Technological Interoperability in Organizations:
## Perspectives and Strategies

Supriya Ghosh
*Arcadia Concepts, USA*

Information Science REFERENCE

**INFORMATION SCIENCE REFERENCE**

Hershey · New York

| | |
|---|---|
| Director of Editorial Content: | Kristin Klinger |
| Senior Managing Editor: | Jamie Snavely |
| Assistant Managing Editor: | Michael Brehm |
| Publishing Assistant: | Sean Woznicki |
| Typesetter: | Kurt Smith |
| Cover Design: | Lisa Tosheff |
| Printed at: | Yurchak Printing Inc. |

Published in the United States of America by
Information Science Reference (an imprint of IGI Global)
701 E. Chocolate Avenue
Hershey PA 17033
Tel: 717-533-8845
Fax: 717-533-8661
E-mail: cust@igi-global.com
Web site: http://www.igi-global.com/reference



Library of Congress Cataloging-in-Publication Data

Ghosh, Supriya, 1964-
Net centricity and technological interoperability in organizations : perspectives and strategies / by Supriya Ghosh.
p. cm.
Includes bibliographical references and index.
Summary: "This book provides understanding on the achievement of interoperability among organizations, focusing on new structural design concepts"--Provided by publisher.
ISBN 978-1-60566-854-3 (hardcover) -- ISBN 978-1-60566-855-0 (ebook) 1. Military telecommunication--United States. 2. Netcentric computing. 3. Internetworking (Telecommunication) 4. United States--Armed Forces--Communication systems. 5. Military art and science--Technological innovations--United States. 6. Military art and science--Data processing. I. Title.
UG593.G48 2010
355.3'3041--dc22

2009034902

British Cataloguing in Publication Data
A Cataloguing in Publication record for this book is available from the British Library.

# Table of Contents

**Section 2:**
**Transformational Perspectives**

*This next section focuses on the transformation aspects of the net-centric future and addresses the book's objectives to provide perspectives and strategies as to how this affects the information enterprise. There are three chapters within this section.*

**Chapter 6**



**Chapter 7**



**Section 3:**
**Configuring the Net-Centric Enterprise**

*This section provides a number of specific topics and enabling technologies that encompass the layered concept of Net Centricity and helps to configure tomorrow's enterprise. There are six chapters within this section.*

**Chapter 8**

# Foreword

Net Centricity allows Military, Civil, Commercial and Personal operations to be executed with the participants asynchronous in time and space. This allows significant improvements in speed, flexibility, and performance over the Industrial Age model where all must be at the factory or office at the same time. The impact is similar to that of the introduction of the automobile into the synchronous world of trains and planes in the transportation field. Achieving these benefits requires significant technological and social changes to be instituted.

As communications bandwidth becomes ever less costly and more widely available, we will be able to not only allow people to process information as they see fit but also allow multiple individuals and organizations to have direct and simultaneous access to information and to each other. We will also be able to support richer interactions between and among individuals.

This book examines these possible advantages and the changes that are required.

*John Stenbit*
*Department of Defense Chief Information Officer/*
*Assistant Secretary of Defense of Networks and Information Integration 2001-2004*

***John Stenbit*** *is nationally recognized for his expertise in developing crucial defense information systems, leading defense transformation initiatives at the Department of Defense (DoD), and applying systems engineering to solve complex problems at large organizations Stenbit was appointed DoD CIO, Assistant Secretary of Defense for Command, Control, Communications, and Intelligence (C3I) by President George W. Bush on August 7, 2001, and then Assistant Secretary of Defense for Networks and Information Integration/Department of Defense Chief Information Officer. He has led the transition to information-centric warfare including creating information systems to support intelligence and operations in Afghanistan and Iraq. For more than three decades, Stenbit held various management positions at TRW Inc., a diversified company that provided high technology products to the space, defense, information systems, and automotive markets. Stenbit holds a BS degree in engineering and an MS degree in electrical engineering from the California Institute of Technology. He has chaired the Science and Technology Advisory Panel to the Director of Central Intelligence, and served as member of the Science Advisory Group to the directors of Naval Intelligence and the Defense Communications Agency. Stenbit currently is a member of the board of a number of different corporate organizations.*

# Preface

I am very pleased to provide all of you with a comprehensive text on the topic of Net Centricity and Interoperability. These are the types of topics that a lot of us hear about, but do not really find any further information, especially within the book stacks of our nearby bookstore.

Before delving into the topic's details, I would like to provide you with thoughts that went into preparing an approach for this book, and thoughts on the upcoming future. This includes an understanding of why this topic is important and the nexus of military and politics. I would like to then provide some notes to you as a reader and a summary discussion of the sections and chapters of this book.

## WHY IS THIS TOPIC IMPORTANT?

Only once in a long time, does our military really transform itself to address fundamental change. Since the dawn of civilization, man has fought wars with each other and a lot of these battles have been recorded in our history books. These history books have described military commands in a hierarchical manner. Battlefield formations have usually been highly structured, with a commander in charge of a division, brigade, battalion, or a regiment. It takes real discipline, composure and a fighting spirit to win wars. The reason for all this structure is that men can then be grouped into action. In our past history, it has taken a massive compilation of physical forces to combat and defeat the enemy.

Well, times are changing. It is no longer necessary to just add up the number of soldiers to figure out the size and power of the military. One person or a small group of people can now be endowed with enough information, intelligence and lethal power to defeat large numbers of soldiers existing in a regiment or a brigade. It is also not necessary to have full command of all information at all times to defeat the enemy. Instead, it is better to provide real-time information and intelligence to the actual decision makers in the battlefield who are trying to actively combat the enemy. Here, it is important to recognize that today's soldiers are more intelligent and more able to take matters into their own hands and thereby – they can save their own lives and the lives of their comrades.

To take advantage of this situation, the military needs to proactively transform. This transformation is fundamental change that is very large in scope. Net Centricity is one word that defines this change. Net Centricity and the ensuing Network Centric Warfare intends to structure the military in a completely different manner to take advantage of today's power of human intelligence and the use of sophisticated technology.

The concept of Net Centricity engulfs within it a lot of new technologies and strategies that have been in the minds of administrators and the military for a long time. The concept includes within it, the ability to conduct incremental change. We all know that fundamental change is difficult, and changing the perspective of our military planners, along with its effects on our citizens takes a long time. This is because people's minds change slowly. This also means that it is important for us to change our political mindset to make a difference as to how our military is administrated and governed.

After the disaster of 9-11-2001, one of the lessons learned that came out of our political consciousness was that our entire government and military infrastructure needs to be more interoperable. The big buzzword was how to "Connect the Dots" properly. The concept of Net Centricity works in tandem with the need for greater interoperability. We realize that if critical information is shared at the particular moment in time when we are all in danger, that it may be possible to avert that danger and eliminate that threat. So in turn, we would like to change our military and defense posture to fully address this need for communicating and networking with each other and within our own organizations.

## Proliferation of Military Technology into Commercial Space

As Mr. Stenbit alluded to in his foreword, net-centric technology provides a military, civilian, commercial and personal component that make us enter into a new era of comprehension and decision making. This is similar in nature to how the Internet has transformed the Information Age in the past decade. Since the Department of Defense created the Internet, the rise of the World Wide Web and the information that it contains may be seen as a first installment of Net Centricity that has germinated within the commercial industry and information technology marketplace.

Just like the proliferation of Internet technology, the net-centric transformation is bringing forth a number of technology opportunities all incorporated within the net-centric enterprise umbrella –this is common for both government and commercial organizations. Challenges such as the implementation of a service-oriented enterprise, achieving effective information security, formulating a common data strategy, or the creation of a common information grid are all being tackled in the commercial market that have had its germination in the military space. This book intends to address this issue head-on, provide an understanding of government and military documents, and point out the appeal to a wider audience who is keen on making use of defense technology into their commercial enterprise.

## The Nexus of Military and Politics

There is another component to this thought process and it is the interaction between the military and political spheres. Although this issue was not addressed directly within this book, it shaped the manner in how information is presented within this book. We all know that just because we come up with a great idea does not mean that this idea will be continually legislated and administered within the government. Net-centric activities within the military greatly depend on the presidential administration, the President's priorities, and Congressional legislation and oversight. The transformation of how our homeland is defended, and how we thwart our enemies overseas depends on Presidential direction, and the resources that we provide to our military and to our soldiers.

It is quite possible that the ongoing net-centric transformative activities will change radically in the future. The name, net-centric, itself may go away as new concepts and strategies take over. However, what will not go away is the current state of our technology, and how we intend to approach our target state. The information within this book is provided in an apolitical third-party manner, so that the advice that is given is more timeless in nature.

Public policy created by Congress and the President may sometimes be reactive, for example, to quell the latest incident that happens in the world. This is also the case for our federal budgetary needs and how the finances are appropriated in Congress. But fundamental change such as Net Centricity takes time, and the fruits of this labor also may not be obvious immediately. So it is expected that a number of the net-centric initiatives within this book will happen in fits and starts. Incremental changes may progress to a point where research and development in technology is finally mature enough to be deployed in the field. We will then all wait as the future is redefined for the new way of doing things.

## NOTES TO OUR READER

This book has been written to take into account all types of audiences. For the uninitiated who wants to simply use the book as a reference on the topic of Net Centricity, the book tries to objectively provide information such that it becomes a handy resource for look-up purposes. This book can be used for the purposes of teaching and as a supplementary text book for classes in military and defense technology.

For subject matter experts in the government and commercial contracting, this book allows you to aggregate a number of disparate topics all under the umbrella of Net Centricity. The book summarizes a lot of information from Department of Defense doctrines and instructions in a manner that is easily disseminated and understandable to everyone.

As a reader, it is best to be fully aware of security and information assurance concerns. If any of the topics discussed in this book is described in enough detail to contain military tactics, then the information sensitivity and the security classification increases. This book only uses information that is labeled as unclassified and for unlimited distribution. Deliberate measures have been taken to stay away from information that can be labeled as For Official Use Only (FOUO) or any information that can ever be labeled as Secret.

The book follows a modular manner where each section follows from the previous one. Each chapter is written in a self-contained manner that delves into a different topic within the same umbrella heading. The focus is on presenting the fundamental concepts, since if you can grasp the concepts and technical ramifications, then it allows you to delve further into tactical details.

Since the net-centric concept is in increments and layers, it is important to provide a perspective to show how it all fits together. Even though technology terms and implementations change over time, the hope is that some of the topics mentioned will stay static long enough, for further research and development from the military to the civilian world.

## HOW IS THIS BOOK ORGANIZED?

The full title of this book, Net Centricity and Technological Interoperability in Organizations – Perspectives and Strategies, is certainly a mouthful. But as the name states, the book tries to provide you with a set of perspectives and strategies as to how the net-centric transformation is taking place, along with tackling the concept of interoperability.

The book has been organized into four sections and seventeen chapters. Each chapter has been written so that they can stand on their own, and each of them introduces new concepts that are distinct in nature. The sections provide a theme as you, the audience read through the book. This allows the book to appeal to multiple types of audiences – from the uninitiated to the subject matter expert who needs to surmise the knowledge base.

## Section I: Becoming Net-Centric

This is the first section that introduces the topic, provides a description of the concepts of Net Centricity and Interoperability, and how organizations can become net-centric.

### Chapter 1: Net Centricity: What Does It Mean?

This chapter provides you with a definition of Net Centricity, discusses terms such as Global Information Grid and Network-Centric Warfare, discusses the Net-Centric Data Strategy and Information Assurance Strategy and provides you with a literature review of current publications.

### Chapter 2: Today's Information Enterprise

This chapter steps back and provides an understanding of today's information enterprise, it discusses information types and terms such as data, information, knowledge and intelligence. It discusses decision making in a collaborative setting and an information sharing strategy.

### Chapter 3: Measures of Interoperability

This chapter focuses on the concept of interoperability, defines the term, discusses specific types of interoperability, and introduces how to measure interoperability levels, and the effects within a distributed enterprise.

### Chapter 4: Net-Centric Operational Environment

This chapter discusses the challenges of today's military, provides a model for a net-centric operational environment and how information should be managed within a theater of operations. It also provides a real-life scenario of an upcoming net-centric environment.

## Section II: Transformational Perspectives

The second section addresses the book's objectives to provide perspectives and strategies by focusing on the transformation aspects, and presents viewpoints from guest authors.

### Chapter 5: Target State for Defense Information Enterprise

This chapter addresses the defense transformation to a target information enterprise, it discusses the vision, goals and objectives, provides a guest author perspective, offers a reference model and how it meets all of the net-centric goals.

### Chapter 6: Net-Centric Military to Civilian Transformation

The chapter specifically provides information on how military to civilian information transfer can take place, describes command-and-control principles of today and how it can be transformed to a net-centric model. It then provides a guest perspective on net-centric computing.

### Chapter 7: Healthcare Transformation in a Net-Centric Environment

This chapter focuses specifically on the healthcare sector, and the transformation of the Military Health System and Veteran's Health Administration. It provides a guest author perspective on military health and discusses the Theater Medical Information Program.

## Section III: Configuring for Net Centricity

This third section provides you with a number of technology topics that all together provide a better understanding of Net Centricity.

### Chapter 8: Use of Enterprise Architecture as a Net-Centric Discipline

This chapter describes the expertise of enterprise architecture; address enterprise architecture planning, implementation, and transitioning to the future state. It then discusses DoD enterprise architecture views and how to meet net-ready key performance parameters.

### Chapter 9: Net-Centric Information Assurance Strategy

This chapter defines information assurance, its basic tenets, and how security controls are implemented, discusses DIACAP procedures, Common Criteria standards, and net-centric information assurance goals.

### Chapter 10: Adhering to Open Technology Standards

This chapter discusses the proliferation of open standards, discusses standards organizations, provides the DoD technology reference model, and then offers a technical standards profile that can be used by any organization.

### Chapter 11: Service-Oriented Architecture and Net-Centricity

This chapter discusses the concept of Service Oriented Architecture, how it is crucial to net-centric principles, provides an understanding of web services, service producers and consumers and the advent of a service-oriented enterprise.

### Chapter 12: Transition to IPv6-Based Networks

This chapter discusses the transition to a IPv6-based next generation Internet, discusses all of the IPv6 specifications, transition to an IPv6 network and a set of IPv6 standards, then it discusses mobile IPv6 strategies.

### Chapter 13: Storage Strategy for the Distributed Enterprise

This chapter discusses storage technologies, how to address digital storage requirements for a distributed enterprise, discusses different enterprise storage design and evaluates network considerations.

## Section IV: Assessing Net Centricity in Organizations

The final section looks at each of the different topics from an assessment perspective. The last chapter then provides an understanding of the upcoming state of Net Centric Service Oriented Enterprise for government and civilian organizations.

### Chapter 14: Architecture Assessment of the Federal Enterprise

This chapter reviews enterprise architecture mandates within the federal government, discusses Clinger-Cohen mandates for information technology investment reform, OMB enterprise architecture assessments, and maturity of organizations.

### Chapter 15: Net-Centric Assessment and Interoperability Testing

This chapter focuses specifically on assessment questionnaire for net-centric and interoperability compliance. This includes data assessment, information assurance, networks and transport mechanisms along with interoperability testing.

### Chapter 16: Technology Evolution Assessment for the Future

This chapter addresses the topic of technology evolution, how to assess organizations with regard to technology standards forecasts, and how to conduct a trade study to acquire hardware and software.

### Chapter 17: Achieving a Net-Centric Service Oriented Enterprise

This last chapter provides the audience with an understanding of the target state after the net-centric transformation. It provides a transition mechanism to get to a net-centric service oriented enterprise, the life cycle processes and how it will benefit different industry sectors.

At the end of the book, there is a set of acronyms and a full glossary of terms to describe each of the nomenclature that have been used throughout the book.

Enjoy,

*Supriya Ghosh*
*June, 2009*

# Section 1
# Becoming Net-Centric

*This is the introductory section that provides an understanding of the concepts of Net Centricity and Interoperability and discusses how organizations can become net-centric. There are four chapters within this section.*

- **Chapter 1:** *Net Centricity – What Does It Mean?*
- **Chapter 2:** *Today's Information Enterprise*
- **Chapter 3:** *Measures of Interoperability*
- **Chapter 4:** *Net-Centric Operational Environment*

# Chapter 1
# Net Centricity:
## What Does It Mean?

## CHAPTER CONTENT

As you explore Chapter 1, it will cover the following topics:

- An Introduction to Net Centricity
- Integrated Approach to Net Centricity
- Understanding Net-Centric Strategy and Goals
- Global Information Grid and Network Centric Warfare
- Issues and Challenges Regarding a Net-Centric Environment
- Literature Review of Net-Centric Publications

## CHAPTER FOCUS

This chapter lets you understand the terms Net-Centric and Net Centricity, and helps you identify the context and paradigm shift this really signifies within the military community and civilian marketplace. Note the integrated approach to implementing Net Centricity, the overall focus of the Net-Centric Data Strategy, the data goals, and concept of the Global Information Grid and Network Centric Warfare. We make sure that we discuss issues and challenges regarding a net-centric environment and provide you with a literature review of commercial and defense publications on the matter.

DOI: 10.4018/978-1-60566-854-3.ch001

## AN INTRODUCTION TO NET CENTRICITY

This book deals with the overall concepts of Net Centricity and the net-centric perspective on the government, military, commercial and personal information awareness. Net Centricity is a borrowed word from the United States Department of Defense and the active military. It translates information superiority into military prowess by effectively linking knowledgeable entities within the battlefield space. However, this book focuses on the military net-centric concept because it is a harbinger to a new way of conducting ourselves within the government and commercial marketplace.

### Fundamental Change of the Current State

The net-centric approach promotes a fundamental change in the way that we conduct our operations. It steps forward from the Industrial Age when we were all tethered to our offices and facilities and were performing work based on stepwise processes and hierarchical management decision making. The new net-centric approach allows asynchronous information assimilation and ad-hoc decision making based on a collective knowledge environment.

The Internet as we currently know it is an early instantiation of the upcoming net-centric transformation. The all-encompassing net-centric approach is a complex concept. Within our book, Net Centricity specifically refers to exploitation of upcoming technologies that provides users the ability to access software applications and Internet-based information services. The expectation is that these applications will be spread over a globally connected information environment that is comprised of interoperable computing and communication components.

In this book, we are using the terms "net-centric" and "Net Centricity" interchangeably. We recognize that in the next decade these terms may take on broader connotations, as these concepts go through a period of adoption. The concepts have traveled beyond the realm of today's military infrastructure, and onto commercial and consumer industries. The term, net-centric can easily be used in the context of the World Wide Web and the Internet, since this is the most well known form of the interconnected, collaborative concept that is at the heart of the term, Net Centricity.

As these terms traverse through the civilian and commercial sectors, the definitions of the terms are changing based on the correct fit and context. A broad definition of net-centric principles would be to:

- Participate in a continuously-evolving, complex community of people and devices
- Provide information and services that are interconnected by a communications network to achieve optimal benefit of resources and better synchronization of events and their consequences.

Although we will focusing a significant portion of this book dealing with the defense and military enterprise, we will be continually be referring to the more broad definition of Net Centricity as it applies to a variety of governments and commercial entities.

### A Military Definition

A strict definition of the term, Net Centricity is provided within the initial Department of Defense (DoD) Net-Centric Data Strategy document published as a DoD Chief Information Officer (CIO) Memorandum (DoD CIO NCDS v1.0, 2003). It states the following:

*"Net Centricity is the realization of a networked environment, including infrastructure, systems, processes, and people that enables a completely different approach to war fighting and business operations. The foundation for Net Centricity is the Defense Department's Global Information Grid (GIG)." ...*

*"The approach to implementing the GIG uses communications, computing, and applications technologies, but also recognizes that the cultural barriers against trust and data sharing must be addressed. To this end, DoD is using a comprehensive, integrated approach to deliver the foundation for Net Centricity."*

From a global military perspective, Net Centricity uses automated sensors, human decision-makers, and technology enablers to achieve:

- Shared Awareness
- Increased Speed of Command
- Higher Tempo of Operations
- Greater Lethality
- Increased Survivability
- A Degree of Self-Synchronization

## INTEGRATED APPROACH TO NET CENTRICITY

Since Net Centricity is not a simple concept, it is best to discuss the topic based on an integrated approach. The DoD Chief Information Officer has developed a comprehensive approach to deliver Net Centricity. This approach provides the guidance for a Net-Centric Data Strategy along with a Net-Centric Information Assurance (IA) strategy that ensures that the information is reliable and can be trusted.

This approach focuses on two very important elements:

- **Data Strategy:** A standardized mechanism to control data elements, definitions, and structures across the enterprise, and gaining consensus on data owners within and across DoD organizations. The focus is to ensure that data administration activities promote Interoperability through the standardization of data elements, minimize duplication of data elements across the DoD, and reduce the need for data element translation.
- **Information Assurance Strategy:** A strategy to ensure that users have confidence that their information has integrity, and authenticity and is available when they need it. The information capabilities need to be robust and resilient in the face of attack, and that the information can be trusted based on the privacy needs of the enterprise.

Following an integrated approach to Net Centricity allows a new paradigm for data management across the defense department, government contracting and the commercial marketplace. The key to the net-centric paradigm is that it recognizes that information that is gathered needs to be usable by unanticipated users and applications as well as for a predefined set of users.

The integrated approach to Net Centricity is implemented in a layered manner. A depiction of this concept (DoD CIO NCDS v1.0, 2003) is shown in Figure 1.

The net-centric implementation can be categorized based on the following layers and named elements within layers, counting from the bottom upwards as depicted in Figure 1.

*Figure 1. Integrated approach for delivering net centricity*

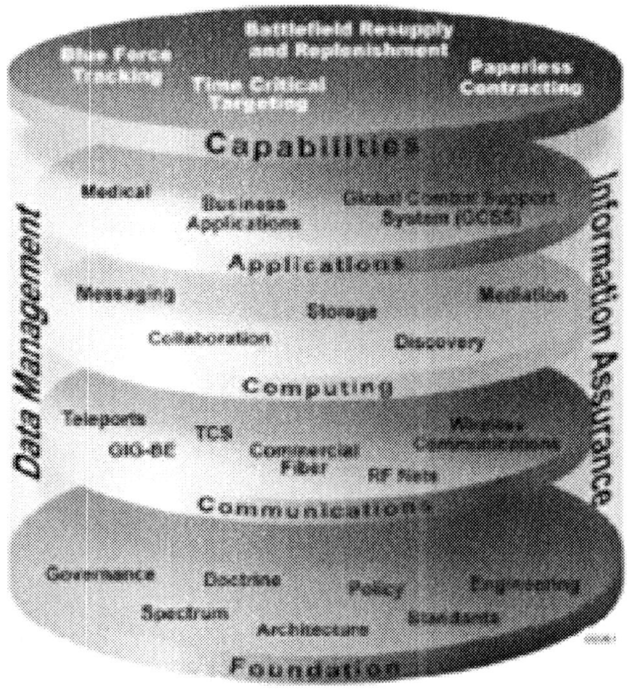


- **First Layer - Foundation**
  - Governance
  - Spectrum
  - Doctrine
  - Policy
  - Architecture
  - Engineering
  - Standards
- **Second Layer - Communications**
  - Teleports
  - TCS
  - GIG-BE
  - RF Nets
  - Wireless Communications
  - Commercial Fiber
- **Third Layer - Computing**
  - Messaging
  - Collaboration
  - Storage
  - Discovery
  - Mediation

- **Fourth Layer - Applications**
  - Medical
  - Business Applications
  - Global Combat Support System (GCSS)
- **Fifth Layer - Capabilities**
  - Blue Force Tracking
  - Time Critical Targeting
  - Battlefield Resupply and Replenishment
  - Paperless Contracting

We start first with the foundation layer that provides the definitions and guidance that organizations need to establish. The second layer focuses on the communications aspects that include specific broadband network and communications initiatives that are ongoing within the DoD. The third layer focuses on the computing infrastructure that includes a set of computer middleware and software technologies and offers organizations the ability to collaborate, send messages and allow data to be discovered and disseminated in a timely manner.

The fourth layer is the applications layer that focuses on developing a full set of business applications such as for medical, DoD business and global combat support. The fifth layer is named for capabilities that take advantage of the previous layers to offer DoD organizations with the power and flexibility to transform current approaches. It includes a set of upcoming operational capabilities such as movement to an all digital, paperless contracting process, or battlefield supply chain management that can be transformed through the use of Net-Centric processes.

## UNDERSTANDING NET-CENTRIC STRATEGY AND GOALS

Net-centric principles are closely related with the interoperability of systems, processes and devices. From a global or enterprise perspective it is best to focus on the visibility and accessibility of data rather than just data standardization within the enterprise. The idea is to identify approaches that improves flexibility in data exchange, and supports interoperability between systems without requiring predefined interfaces between two exchanging entities. This flexibility is essential in the "many-to-many" exchanges of a net-centric environment.

Net Centricity does not eliminate the need for tightly engineered, predefined interfaces between systems. Instead the objective is to increase the potential for many other systems to leverage the same data without having to anticipate the use of this information within the system's development cycle. For example, real-time systems that have been engineered as a tightly-defined set of interfaces can offer "exposure" services that work behind the scenes. The effort is to collect real-time data, store it, and provide data access and data discovery through an enterprise-level interface.

However, in an environment where applications and information technology systems are continually being developed, deployed, migrated, and replaced, making allowances for unanticipated systems or interfaces is essential. As part of the Net-Centric Data Strategy, exposure to services can be designed so that they have little or no effect on day-to-day operational performance. However, they may still provide access to unanticipated users and provide the operational data to other critical systems and processes based on the number of affected stakeholders.

*Figure 2. Overall Focus of Net-Centric Data Strategy*

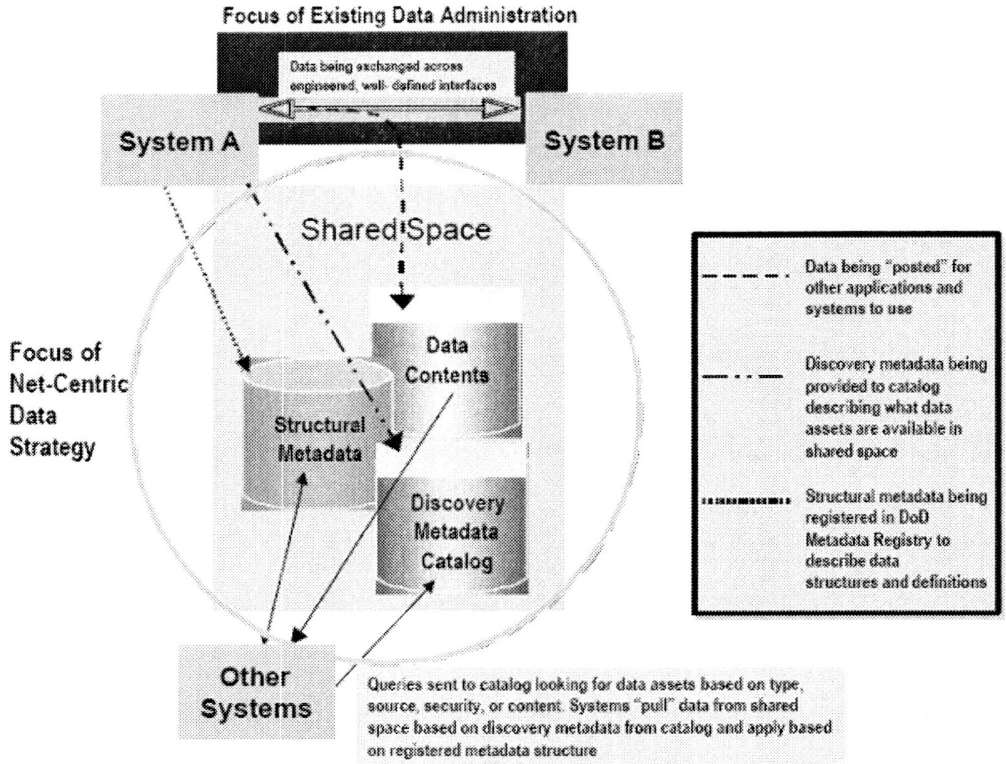


Figure 2 provides an illustration of the overall focus of the Net-Centric Data Strategy (DoD CIO NCDS v1.0, 2003) that takes into account the expanded concept of data entities that share information across the enterprise. Note that the top part of the diagram presents System A and System B that is sharing data based on a pre-defined set of data interfaces. The Net-Centric strategy then asks System A to place data elements within an enterprise-level "Structured Metadata" database and an enterprise-level "Metadata Catalog" that can be then disseminated and shared with other systems.

Organizations can query the enterprise data catalog and look for data assets based on data type, data source, and data content. Systems can then "pull" data from the shared data catalog, and understand the content of the data based on a registered metadata structure.

The direct benefits are to improve flexibility in data exchange between organizations, support interoperability between systems and eliminate the necessity for predefined, pair-wise system interfaces. This flexibility then allows the "many-to-many" exchanges to happen simultaneously and in an automated manner.

This illustration recognizes the existence of System A and System B that is exchanging data based on predefined interfaces. These systems may even be termed as "legacy" over time. New systems can then be more easily consumed within this strategy, and the strategy allows other potential systems to be leveraged and accessed in a timely manner.

## Data Strategy Goals

Here are two goals that are central to the Net-Centric Data Strategy:

1. Increasing the data that is available to communities or the Enterprise
2. Ensuring that data is usable by both anticipated and unanticipated users and applications

The data and information assets pertain to all types of electronic content that is collected and disseminated. This includes all legacy and new data assets such as:

- System files
- Databases
- Electronic Documents
- Official electronic records
- Graphic images and photos
- Audio files
- Web sites and hyperlinked sites and Web Logs
- Data access services

Here are a set of desired objectives for promoting a net-centric environment. These goals align with the DoD CIO guidance provided within defense doctrines and instructions.

- **Make Data Visible** – Public and global data can be discovered by users and applications based on data catalogs, registries and search services. All data assets can be are advertised based on providing metadata which describes the assets.
- **Make Data Accessible** – Users and applications can post or share data based on descriptive information that is provided to a data catalog and which is visible to the entire enterprise.
- **Institutionalize Data Management** – The data owners and groups of users incorporate data approaches that are codified into processes and practices. This allows governance based on best practices and sustained leadership and allows training of new users.
- **Enable Data to be Understandable** – The strategy needs to allow users and applications to comprehend the data both structurally and semantically so that the data may be readily used for specific needs.
- **Enable Data to be Trusted** – Users and applications need to be able to determine the authority of the data, based on the data source, its pedigree, the security level, and access role for the data asset.
- **Support Data Interoperability** – Allow many to many exchanges between data to occur among systems and through their interfaces, that may sometimes be pre-determined and sometimes to be unanticipated. This type of exchange requires the preparation of metadata so that we allow mediation or translation of data between interfaces.
- **Be Responsive to User Needs** – The different perspectives of end users and data consumers need to be incorporated within the overall approach so that there is a continual feedback mechanism to ensure satisfaction.

### Information Assurance Strategy Goals

As part of the net-centric transformation, we need to recognize that information assurance is a key element for progressing toward the future. Information assurance goals are to assure that within the overall enterprise, the information assets, information systems and the underlying infrastructure are available, can be trusted, and have a degree of authenticity.

For the upcoming future, it is important to ensure that information assurance addresses the elements of technology, operations, processes and people. The desired set of objectives (DoD IA Strategic Plan, 2008) can be used to address the following goals for your enterprise.

- **Protect Information** – The enterprise needs to safeguard data as it is being collected, analyzed, processed and disseminated. This allows us to ensure that all information has a level of trust commensurate with the mission needs.
- **Defend Systems and Networks** – The enterprise needs to recognize, react and respond to threats, vulnerabilities and deficiencies by ensuring that all systems and networks are capable of self-defense.
- **Align Network Mission Assurance through Integrated IA Situational Awareness and IA Command and Control** – This means that the enterprise integrates their user-defined operational picture and this is synchronized with network operations. Within the DoD this allow the military to have a common operating picture to conduct net-centric warfare.
- **Transform and Enable IA Capabilities** – It is best to enable IA capabilities through innovation and experimentation that leverages emerging technologies and implements best practices. This allows you to refine the assurance processes and improve cycle time, reduce risk exposure and increase your return on investments.
- **Create an IA Empowered Workforce** – Since information assurance is still a new endeavor, it is best to equip your people so that they can support the changing demands of the information assurance and technology enterprise.

## GLOBAL INFORMATION GRID AND NETWORK CENTRIC WARFARE

In this section, we move on to define two new terms that are closely linked with the net-centric vision and data and information assurance strategy. In implementing the net-centric goals and processes, it is important that our military implements it within its overall communication and application infrastructure. This then allows us to define the concepts of the Global Information Grid (GIG) and Network Centric Warfare (NCW).

The GIG is a complex term that may mean a number of thoughts based on an individual's perspective. A broad definition of the term based on guidance documents (GIG CRD, 2001) states that it encompasses the following:

- GIG is defined as the globally interconnected, end-to-end set of information capabilities, associated processes, and personnel for collecting, processing, storing, disseminating, and managing information on demand to warfighters, policymakers, and support personnel.

- GIG includes all owned and leased communications and computing systems and services, software (including applications), system data, security services, and other associated services necessary to achieve information superiority for the United States military.
- GIG is the physical manifestation of the doctrine of Network Centric Warfare.

DoD component organizations such as the Defense Information Systems Agency (DISA) who is the executive agent for GIG implementation provides an ongoing definition of the GIG as the infrastructure is built out and implemented. However, it should be noted that because of information assurance necessities, most of the manifestations of the GIG is classified as Secret or higher. This means that information is compartmentalized at higher levels of security based on a graded "need to know" basis. Information assurance plays an important role in to ensure that the exact strategy and tactics are classified based on the needs of the users and the affected systems.

Network Centric Warfare (NCW) is a military implementation of Net Centricity and it represents a fundamental shift in military culture. It provides a vision away from a compartmentalized war scenario and toward interconnected units that operate cohesively. In the DoD NCW Report to the Congress (Wilson, C., 2005) the overall goals of NCW are mentioned as follows:

- A robustly networked force improves information sharing
- Information sharing enhances the quality of information and the ability for shared situational awareness
- Shared situational awareness enables collaboration and self-synchronization and enhances sustainability and speed of command
- The above goals thereby increase mission effectiveness

A broad definition of Network Centric Warfare can be written based on the following thoughts and judgments:

- NCW uses computers and communications to link people through information flows that depend on the interoperability of systems used by all U.S. armed forces.
- NCW involves collaboration and sharing of information to ensure that all appropriate assets can be quickly brought to bear by commanders during combat operations.
- The network centric approach to warfare is the military embodiment of information age concepts. Studies have shown that networking enables forces to undertake a different range of missions than non-networked forces, by improving both efficiency and effectiveness of operations.
- Procurement policy to support NCW is also intended to improve economic efficiency by eliminating stove-pipe systems, parochial interests, redundant and non-interoperable systems, and by optimizing capital planning investments for present and future information technology systems.

Based on this broad definition, NCW is closely aligned with the net-centric vision, and can be stated based on the following objectives:

- The use of self-synchronization that allows military personnel to do what needs to be done without receiving traditional orders
- An improved understanding of higher command's intent

- An improved understanding of the operational situation at all levels of command
- An increased ability to tap into the collective knowledge of all U.S. (and coalition) forces to increase the instantaneous delivery of information necessary for fighting

We do not want to espouse that modern military environments have become far too complex to be understood by any one individual, organization, or even military service. This has direct effects as to how wars are fought and won. However, as tomorrow's battlefield and theater changes, the idea is to adopt the principles of NCW.

In this case, NCW provides more capabilities to edge entities, such as theater command systems and applications so that they can conduct the military missions themselves. This means that they do not have to rely on a centralized command-and-control infrastructure to provide them with information on how individual battles should be fought. Military personnel are urged to pull information from distributed repositories, rather than having a centralized command attempt to anticipate their information needs and push it to them.

## ISSUES AND CHALLENGES REGARDING A NET-CENTRIC ENVIRONMENT

As we establish the need for a net-centric transformation, we want to recognize that there are a great many issues and challenges that need to be worked out. As we navigate through this paradigm shift, we also do not want this thought process to move from the military to the civilian marketplace and be seen as a panacea to our current information-related problems and issues.

As we research various technologies, we can say that all of the net-centric strategies and tactics may or may not prove to be useful. It is up to individual decision makers to make this determination. The DoD has stated that their component organizations must continue to refine the rules and theory of network centric warfare through simulation, testing, and experimentation.

### Challenges in Information Reliability

One of the biggest challenges in establishing a proper net-centric information environment is to ensure that your information set within the enterprise is reliable and trustable. Setting up a basis for trust, with regard to information sharing presents one of the greatest enterprise challenge. Trust is necessary in all aspects of an information system. This includes trust in the system's availability, the participants' identities, and the data's dependability and integrity.

We all know that most organizations today are grappling with information firewalls and software patches that attempt to keep intruders out and keep their data safe. Since in a net-centric information environment, the data is being used by multiple communities, and this exacerbates the problems. Critical information that is trusted and assured also needs to be kept secure throughout the entire information's useful life span.

### Challenges in Network Centric Warfare

A net-centric information environment may increase certain advantages in warfare, but it may also indicate that relying on information systems can sometimes lead to unexpected results. A review of information-age warfare has shown that this new environment is increasingly path-dependent. This means that small changes in the initial conditions may result in enormous changes in outcomes. Speed is an important characteristic for NCW because it enables a military force to define initial conditions favorable to their interests, and then pursue a goal of developing high rates of change that an adversary cannot outpace.

A number of thought leaders and military planners have noted that military decisions may also not always lend itself to information-based rational analysis. Most of the DoD community that includes the military services, national security establishment and Washington policy makers have found that it is not easy to make decisions based on a data-dependent military doctrine. Issues that have been raised by thought leaders that have been reported to the U.S. Congress as part of the NCW doctrine (Wilson, C., 2005), include the following:

- Information flows may be governed by a diminishing marginal utility for added effectiveness. Quantitative changes in information and analysis may lead to qualitative changes in individual and organizational behavior that are sometimes counter-productive.
- An information-rich, opportunity-rich environment may shift the value of the information, redefine the mission objectives, and possibly increase the chances for perverse consequences.
- Reliance on sophisticated information systems may lead to management overconfidence.
- Different analytical interpretations of data may lead to disagreements among commanders about who is best situated to interpret events and act on them.

However, as more progress is made towards a net-centric information environment, it is clear that using the latest technology tools is certainly beneficial and may lead to the desired outcome more quickly. For example, when digital information is sent through data links that can then be shared and processed instantaneously – this produces a significant advantage over other organizations that primarily rely on voice-only communication.

The strategy to move toward a net-centric information environment is to address the challenges of uncertainty and risk by focusing primarily on data. Instead of relying on proprietary applications and programs, users are urged to post all data assets within the shared to a common information enterprise. Instead of pushing information out based on pre-determined interfaces, the strategy allows users at any level to both take what they need and contribute to what they already know.

## Creation of a Community of Interest (COI)

The defense community has introduced the concept of a Community of Interest (COI). A COI is a collaborative group of users who need a common set of information, and must have a shared vocabulary to exchange their information. COIs can work toward data characteristics and content that will be tagged in an agreed-to manner. A COI can range from pre-established groups with ongoing arrangements to unanticipated users and non-traditional partnerships that develop on an ad-hoc basis.

Users within a COI, based on their authenticated identity and authorized access role can then view, access and edit the data in a common manner. The goal is then to make the modified data visible, accessible and understandable within a group setting.

## LITERATURE REVIEW OF NET-CENTRIC PUBLICATIONS

This section provides you with a listing of documents that provides a view of Net Centricity, net-centric policy and outcomes. Although each of the chapters of this book has an associated set of references, this list provides you with key commercial and defense publications that are central to the understanding of the net-centric transformation.

### Commercial Publications

1. Alberts, D.S., Hayes, R.E. (2003) *Power to the Edge*, CCRP Publication Series
   - This book refers to Net Centricity and the ability of giving power to the edge based on achieving command and control agility, and increasing the speed of command over a robust, networked grid.
2. Alberts, D.S., Garstka, J.J., & Stein, F. P. (1999). *Network Centric Warfare: Developing and Leveraging Information Superiority*. CCRP Publication Series.
   - This is a follow-up book on conducting Network Centric Warfare based on the concepts of giving power to the edge in the battlefield and theater.
3. Morris, E., Levine, L., Meyers, C., Place, P., & Plakosh, D. (2004). *System of Systems Interoperability (SOSI): Final Report*, CMU Technical Report
   - This provides a well thought out review of the concept of Interoperability, its various definitions and how it affects large organizations.
4. Starr, S. H. (2004) *The Challenges Associated with Achieving Interoperability in Support of Net-Centric Operations*, White Paper, Barcroft Research Institute.
   - This paper provides an overview of the topic of Net-Centric operations and discusses how interoperability measures are necessary to achieving this capability.
5. Ghosh, S., (2007) *Promoting Net Centricity through the use of Enterprise Architecture*, Handbook of Enterprise Systems Architecture in Practice, IGI Global.
   - This chapter in a book correlates how net-centric capabilities can be achieved based on following proper enterprise architecture planning to fully document an organization's enterprise.

### Defense Publications

6. DoD CIO (2003). *DoD Net-Centric Data Strategy*
   - The Net-Centric Data Strategy is a key enabler of the agency's transformation that establishes the foundation for managing data in a net-centric environment.
7. DoD CIO (May 2007). *DoD Net-Centric Services Strategy*
   - The Net-Centric Services Strategy expands upon the Net-Centric Data Strategy by connecting Services to the Data Strategy goals and establishing a net-centric environment that leverages shared services and Service Oriented Architecture.
8. DoD CIO (2004). *DoD NII Net-Centric Checklist*
   - The purpose of the Net-Centric Checklist is to assist program managers in understanding the net-centric attributes that their programs need to implement to move into the net-centric environment within the Global Information Grid.

9. DoD CIO (2007). *DoD Information Sharing Strategy*
   - The strategy establishes the vision and goals for information sharing, while paving the way for a more detailed initiative to document the implementing actions necessary to achieve these goals and realize the vision of Net-Centric Operations.
10. DoD CIO (2002). *DoD Information Assurance Policies*
    - This directive establishes policy and assigns responsibilities to achieve DoD information assurance through a defense-in-depth approach that integrates the capabilities of personnel, operations, and technology, and supports the evolution to network centric warfare.
11. DoD CIO (2008). *DoD Information Assurance Strategic Plan*
    - This plan set relevant objectives and the actions critical to securing the Net-Centric Global Information Grid (GIG) and achieve the long-term vision based on a trusted information base.
12. Wilson, C. (2004). *Network Centric Warfare: Creating a Decisive Warfighting Advantage*. CRS Report to U.S. Congress.
    - This document describes DoD strategy for implementing network centric warfare and was presented to the U.S. Congress. Key elements for NCW implementation include refining the rules of NCW through simulation, testing, experimentation, and combat experience.

## REVIEW OF CHAPTER GOALS

The goals of this chapter were to address:

1. Introduce the concept of Net Centricity:
   - Why is it relevant, what is the motive behind it, how can it allow us to transform our defense posture?
2. Provide an integrated approach to Net Centricity:
   - What are primary concept layers, how is it all integrated together?
3. Discuss the fundamentals of the Net-Centric Data Strategy:
   - How is this new data strategy different from the past, and how will it affect information systems?
4. Discuss the DoD Net-Centric Data Goals:
   - This information provides goals set forth by the DoD for all its component organizations.
5. Define the concepts of Global Information Grid and Network Centric Warfare:
   - How are these complex concepts defined and how do they work together?
6. Discuss issues and challenges for the upcoming net-centric environment:
   - State challenges in harnessing new technology and also monitor the issues in implementing this new environment for the future.
7. Review of Defense Publications on Net Centricity:
   - An annotated bibliography of a number of primary documents that helped spell out the contents of this chapter.

## REFERENCES

Alberts, D. S., & Hayes, R. E. (2005, April). *Power to the Edge*. Command and Control Research Program Publication Series. Available at www.dodccrp.org

Department of Defense Joint Publication 1-02 (2001). *DoD Dictionary of Military and Associated Terms*. Washington, DC: Author. Retrieved from http://www.dtic.mil/doctrine/jel/new_pubs/jp1_02.pdf

Ghosh, S., (2005, April). *Summary of DoD Net-centric Assessment of TMIP Block 2 System*. DoD OASD(HA), Joint Medical Information Systems Program Executive Officer. Alberts, D.S., Garstka, J.J., & Stein, F. P. (1999). *Network Centric Warfare: Developing and Leveraging Information Superiority* (2nd Ed.). Washington, DC: CCRP Publication Series.

Ghosh, S. (2007). Promoting Net Centricity through the use of Enterprise Architecture. In P. Saha (Ed.), *Handbook of Enterprise Systems Architecture in Practice* (pp. 253-270). Hershey, PA: IGI Global.

Office of the Assistant Secretary of Defense for Networks and Information Integration. Department of Defense Chief Information Officer, (2001). *Global Information Grid (GIG) Capstone Requirements Document (CRD), Flag Draft*. Retrieved from http://www.dfas.mil/technology/pal/regs/gigcrdflag-levelreview.pdf

Office of the Assistant Secretary of Defense for Networks and Information Integration, Department of Defense Chief Information Officer Directive 8320.02-G (2006, April). *Guidance for Implementing Net-Centric Data Sharing.*

Office of the Assistant Secretary of Defense for Networks and Information Integration, Department of Defense Chief Information Officer (2008). *Department of Defense Information Management/ Information Technology Strategic Plan 2008-2009.*

Office of the Assistant Secretary of Defense for Networks and Information Integration, Department of Defense Chief Information Officer. (2003). *DoD Net-Centric Data Strategy* (Version 1.0).

Wilson, C. (2005, March). *Network Centric Warfare: Background and Oversight Issues for Congress*. CRS Report for Congress.

# Chapter 2
# Today's Information Enterprise

## CHAPTER CONTENT

As you explore Chapter 2, it will cover the following topics:

- A Glimpse at Today's Information Enterprise
- Structuring the Information Enterprise
- Information Functions within the Enterprise
- Understanding the Decision Making Pyramid
- Information Sharing Strategy
- Tomorrow's Enterprise – Future Trends

## CHAPTER FOCUS

This chapter shifts you from the complex topic of network centric warfare to address the basic fundamentals of today's information enterprise environment. It addresses how our information enterprise is structured and what forms of information forms are there. This includes use of text, images, voice, audio/video and data. It then provides an understanding of the information functions as to how to use these information forms. It helps create, display, store, process and transport information, and work on how these converge. It then introduces the concept of a decision pyramid which in detail explains data, information, knowledge, intelligence and how decision making requires all of them. It further discusses an information sharing strategy for large organizations and the information sharing value chain. The chapter ends by discussing a number of tomorrow's information technology future trends.

DOI: 10.4018/978-1-60566-854-3.ch002

## A GLIMPSE AT TODAY'S INFORMATION ENTERPRISE

Everyone says that we live in an "Information Age". With the beginning of the 21st century, it has become highly important to talk about the entire information enterprise. For large government and commercial organizations, this is a vast stage where information over digital and electronic networks is expanding at a very large rate.

Given that today's enterprise is flooded with more information that can ever be processed, how can the special circumstances of today's information age be turned into an organizational advantage? The decentralized nature of today's information enterprise is at the heart of the Net-centric paradigm. How individuals, groups, and large departments handle the navigation and communication of large amounts of information is the key to addressing Net Centricity and its complex infrastructure.

We recognize that the information industry today is in the stages of a complete transformation. Historically, the industry has been organized on the basis of the form of information. Businesses have addressed single form of information provided as text, voice, image, audio, video and data. However, in the most recent decades there has been great movement in digital electronics to allow a complex multimedia of information that is a combination of text, images, audio or video.

This advent of multimedia information has been coupled with fast growth of digital networks that have expanded the ability to carry information from previous narrow band to newer broadband communications technology. Now that digital electronics is fast becoming the dominant technology for every information intensive organization, a realignment of our daily operational activities is inevitable.

The information industry is starting to reorganize themselves into the following three broad groups:

- Organizations that are involved in the creation and collection of diverse types of information
- Organizations that are involved in the manufacture of a wide variety of information appliances and their requisite user-based applications
- Organizations that are forming to engage in the various modes of information transport, along with routing and dissemination

This rapid movement of the information industry is enabled by the computer hardware industry, which is rapidly improving the logic and processing capability of computer chips, the ability for personal computing devices to connect to an all-purpose network, and the insatiable necessity for storage space, that allows the capacity to store information in large digital repositories.

The associated computer software industry provides the enabling technology for people to engage in information gathering, content management, instant mail and communication, dissemination of large documents, and navigating the World Wide Web. It is understood that the information industry is far from being a mature industry, since the global penetration of digital communications is still fairly low. However, that percentage will surely increase as countries throughout the world participate in fueling this unprecedented growth.

*Table 1. Core competencies of the information enterprise*

| Types of Information Business | Core Competencies |
|---|---|
| **Information Content** | Fostering and managing creativity, information gathering and application programming skills |
| **Information Appliances** | Global presence, strategic sourcing, design and manufacturing expertise, concurrent engineering and public awareness, information portability and scalability |
| **Information Transport** | Network management, interactive communications, billing systems, and IT infrastructure platforms |

## STRUCTURING THE INFORMATION ENTERPRISE

In this section we describe the core competencies of the information enterprise and the different forms of information to include text, images, voice, audio/video and data. As we view information technology move towards electronics and digitization, it is apparent that the information industry is undergoing a structural transformation. The International Engineering Consortium (2001) has written extensively about this topic within its tutorials, and we have used their concepts throughout this chapter.

The information enterprise is driven by the definition and extension of certain core competencies. The way to describe a critical core competence would be to address the mastery of enabling technologies that helps produce a host of information products. As we review how the information enterprise is structured, we recognize that the information industry can be viewed as consisting of three different elements:

- **Information Processing** – which deals with the information creation, update, content distribution and storage of information
- **Office Technologies** – which provides the devices and tools such as the personal computer, operating system, word-processing software, and the ability to automate information tasks
- **Telecommunications** – which provides the capacity for information to flow from one area to another, the data networks and physical communication lines

However, based on the rapid technological convergence in the last decade, these three elements have been merging based on a common, distributed infrastructure. These can be represented as core competencies as shown in Table 1.

### Forms of Information

In reviewing the information enterprise as a whole, it is best to provide a broader definition of all the forms of information. Information can be provided in five basic forms: voice, text, images, audio/video and data. Each form of information has been represented at the retail level by one or two industries. For example, imaging consists of the use of cameras, industrial imaging, Xeroxing, faxed documents, and digital drawing tools. Video information has been used by the entertainment industry along with television, satellite feeds and video communication.

The information processing aspects of each of these information forms includes how information is created, displayed, stored, processed and distributed. Here is a synopsis of each of the information forms.

## Use of Text

This is the most basic form of information communication and dates back to our collective history. Back in 1455, based on the invention of the printing press, the use of printed text started the original information industry. Throughout the 19th and 20th century, the use of printed text allowed newspapers and magazines to flourish, along with the ability for people to communicate based on the preparation and printing of books. The primary focus of the text-based industry has been to create content and gather information printed in books and libraries. The secondary focus has been to address the need for easy dissemination and distribution of the written text material. The primary technology that enabled this effort was the electromechanical printing press, offset type, and mechanical foundries. In the past two decades, the word processing software application on a personal computer has replaced the typewriter as the primary mechanism for creating and distributing text-based information. Then with the advent of the World Wide Web and e-mail, the use of text has expanded greatly and is currently the primary form of communication.

## Use of Images

We are all familiar with the famous saying which goes, "a picture is worth a thousand words". The use of images as a means of communication started back in the 1830s when men were able to use chemical photo-developing techniques to capture images from a camera. The photography industry has led in the capture, development and dissemination of images. Images have also been used in newspapers as drawing cartoons and sketching of images as comic book characters have extended the grasp of imagination. At this time, the imaging industry is a large and varied group that includes medical imaging taken in the form x-rays and CT scans, to Hollywood movie characters depicted as a publicity stunt. The large industry that grew out of Xerox copiers and facsimile machines has added to the necessity to store graphical images in analog or digital form. Images are now being rendered as a digital reproduction and stored as computer files. Images also are being sorted based on textual data that is defining what is contained within the image file.

## Use of Voice

The telephone industry created by Alexander Graham Bell back in the 1870s became the predominant form of voice communication. Telephone networks connected across the country allowed people far away to "reach out and touch someone" based on their picking up a telephone and calling the person. In the 1930s, voice communications was enhanced by the introduction of the "talkies", the movie industry providing both a live picture and a voice recording. The telephone industry has been a mainstay that has expanded into telecommunications that not only includes voice, but also data, and facsimile-based communications. In the recent decades, voice communications has expanded to include wireless cellular communications and the use of personal cell phones that can be carried anywhere. Within short distances, there have also been walkie-talkies and the use of two-way communication devices that allow burst of voice communications via radio signals. Voice communications is now being digitized so that the noise ratio for sending voice signals through the air is minimized, and large digital switches can efficiently transmit the voice information to long distances.

## Use of Audio/Video

This type of information communications comprises of audio information such as music along with video such as motion pictures. Audio and video industry has been the mainstay of the entertainment industry, since music, movie theaters and television networks all provide the greater public with entertainment choices. However, audio and video information is clearly necessary in all types of contexts, including today's military. Satellite-based communications provide long distance audio and video signals for any television monitor, and can capture enemy reconnaissance along with command and control information. Audio and video, for example, is also being used for tele-medicine and tele-health purposes for ensuring the health of a warfighter. Cable television networks along with satellite-based entertainment providers currently are providing a wide variety of audio and video signals to the general public. Audio and video are certainly the most animated form of information communications and provides a lot of information in a short period of time. As digital forms of audio and video expand across the globe, it provides the most effective form of communications that can reach people in distant parts of the world.

## Use of Data

Data is being separated from each of the other information forms as the newest form of communications. With the advent of computer, it is recognized that communications has gone beyond mere human-to-human interactions. Computers require automated communications, where data is sent back and forth between systems based on automated handshake and protocols. The industry has grown from mainframe computers to personal computers, networked devices, and wireless devices. Computing has required information processing and storage, based on data sent from one location to another. It is also recognized that "metadata" is also important to capture and send, since metadata is defined as data that describes data. Data is primarily text-based, but includes symbols and software notation that only computers can understand. Information communications across the local area or across a wide area focuses on data communications and is customized to send data at very fast speeds for long distances. As networks are expanded to broadband connections, it allows ever greater amounts of data to be transmitted between organizations, including the government and military.

# INFORMATION FUNCTIONS WITHIN THE ENTERPRISE

Now that the chapter has discussed the various forms of information within the enterprise, it is important to evaluate how these information forms can be managed. The primary operational functions for depicting information are presented in Table 2 below.

The information management functions can be summarized as follows:

- **Creation** – Information needs to be created, collected, updated, and prepared. As content is created in digital form, it is easier to repurpose and manipulate. This creation of content would include all the forms, text, images, data, voice, and audio and video. Content can be created by any type of user that includes the general office worker who is writing an e-mail, to persons in the field creating instant messages. Audio and video require greater effort and a larger talent pool, which can then be prepared for a greater audience.

*Table 2. Examples of information functions*

| | Text | Voice | Image | Audio/Video | Data |
|---|---|---|---|---|---|
| **Content Creation and Collection** | Writers, authors, journalists, office workers | Telephone operators, directory publishing, voice recording | Photographers, cartoonists, graphic artists, medical imaging providers | Musicians, movie producers, videographers, production studios | Website developers, automated computer software, data processing |
| **Display Formats and Devices** | Books, magazines, web sites, newspapers, e-mails | Telephone equipment, cell phones, personal communication devices | Photographs, slide projectors, fax machines, printers, scanners, copiers, | Television, movies, radio, stereo equipment, movie projectors, DVDs, videos | Personal computers, wireless hand-held, laptop, terminals, workstations |
| **Storage and Memory Devices** | Libraries and information services | Pagers, answering machine and voice-mail | Image files, pictures, slides, three dimensional rendering | Audio CD, video DVD, blue-ray disks and game cartridges | Computer disks, network directories, tapes, optical disks |
| **Processing and Applications** | Printing and publishing, word processing, hypertext linking | Voice processing and tele-communications networking equipment | Photo developing services, image processing | Audio production, movie and video production, editing, | Personal computing applications, wireless data applications, database transaction processing |
| **Distribute and Transport** | Postal mail delivery, electronic mail, fax, instant messaging | Telecom voice service, communications service, paging, messaging | High-speed image transfer, medical imaging device transfer, sending DVDs | Television cable, satellite-based delivery, video purchase and rental, | LAN, WAN, distributed data networking, IP-based delivery |

- **Display -** Historically, each form of communication has had its own types of display. This means that text has primarily been on paper, images have been based on photographs, voice has been based on the telephone, audio and video has been based on recordings and movies, and data has been through computing or data devices. This entire paradigm shift is based on all of the forms of communication being digitized. Now, all of the information forms can be stored as digital files and displayed together as necessary based on end user needs and demands.
- **Storage –** Since the creation of information and the consumption of information is separated out, and because information tends to have lasting value, it is very important to store information on a long-term basis. Digital information stored within computers can either be stored directly on hard disk drives, or on transportable media such as tape, disks, or on network drives. The key driver to the value of a storage system is how easily it is accessible to the user. Hence, storage devices that allow easy sorting and searching of the data is preferable, especially when the storage devices are distributed geographically across large distances and within disparate networks.
- **Processing –** The processing of information is required to manipulate data based on intelligence and decision making. A variety of processing approaches have been taken based on the information form. For text information, publishers have used word processing, web-based content management and desktop publishing. In the imaging business, back-end image labs along with digital processing have been used to format the image to user's satisfaction. For audio/video, it is necessary to use editing software applications and post-production techniques to edit information to the vision of the director and movie producer. Data processing has been common for maintaining and updating databases that store individual strings of data in fields and rows.
- **Transport –** This function allows information to be distributed and transported wherever it is necessary. A large information infrastructure is necessary to communicate and transport each of

the information forms. The telecommunications industry provides a vast wiring mechanism that includes copper and fiber-optic wires to travel across great distances. This has now been extended to wireless networks based on cellular and radio communications. Satellite communications have provided point-to-point delivery based on antenna and receivers at distributed locations. Each of the forms requires a different capacity of transmission based on the size of the digital files. Text files are usually very small, while audio and video files can be very large. Recently, transmission networks have been upgraded to broad-band communications, to keep up with rapid need for sending and receiving and variety of information in different forms.

### The Convergence of Information

As has been stated in the previous section, the information enterprise is undergoing rapid changes based on fast moving technology upheaval. Each information form that was previously managed via "analog" means is fast becoming all-digital. This means that information is being transformed via applied mathematics to digits of ones and zeros. This is the process of digitization that outputs all information forms into digital computer files. Based on common file formats, text files can be stored as .txt, while data files can be stored as .dat, audio files as .avi, images files as JPEG and movie files as MPEG files. As information is all stored on the computer as files, this causes a fast convergence of all information into a single computer directory.

Historically, information was created, stored, manipulated, and transformed based on its form. Textual information mostly remained as hard copies, based on handwriting, or the printed paper. Audio information was stored as analog waveforms, captured on vinyl or magnetic tape. Still and moving images were rendered in chemical form. Moving images became stored in analog form based on videotape technology. However, data always has been in digital form or else it would not be understood by computers.

The digital world is characterized internally by sameness since all computer bits and bytes are represented by a constellation of 1 and 0. As more and more information is digitized, it loses its past and all become very similar in nature. This causes new problems such as authenticity and provenance of the information, which is very hard to figure out in a digital world. Data forms have to be characterized by metadata, which provides critical pieces of information to determine the nature of the information form being stored.

It is important at this time to recognize this digital convergence, since it is the basis for our entire information infrastructure. Digital information is essentially distortion-free, easily replicable and easy to transport through large distances. It is also more difficult to maintain intellectual property rights of digital data since it can be easily repurposed by users for different purposes.

Convergence of multimedia data combines text, images, audio and video, and it gives the creator, the editor, or the publisher greater power to mold information to their desires. Multimedia data is what we find currently over the World Wide Web, and it's the choice of most end customers.

## UNDERSTANDING THE DECISION MAKING PYRAMID

As we explore today's information enterprise, it is helpful to provide an informed perspective of its essence. Words that we all use in common can frequently take on more substantive meaning, as we explore

*Figure 1. The decision making pyramid*

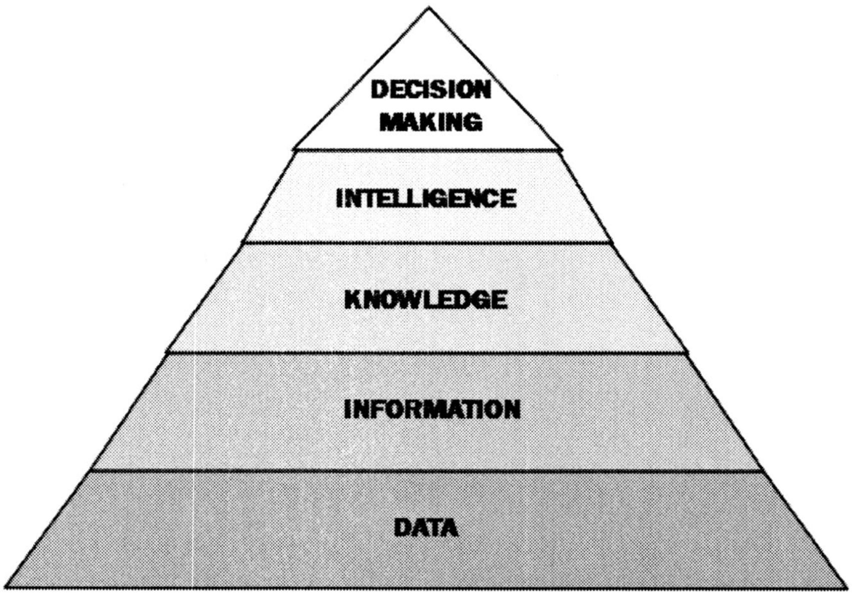


the context of our surrounding. Since the net-centric transformation is a complex endeavor, it is best to address the information enterprise from the broadest context.

Information is proliferating in greater parts of our society and throughout the globe. So it becomes more and more difficult to make an informed decision until all of the facts and associated information is analyzed. Decision making is difficult when one does not understand their surrounding and the context as to how, when, where and how a decision is made.

Figure 1 below provides a perspective to decision making in the form of a pyramid. The idea of this pyramid is simple. As you go upwards in this pyramid ladder, the amount of information lessens, but the amount of understanding increases.

Each rung of the Decision Making Pyramid sits on top of another and is incremental in its ability to collect, disseminate, update, surmise and comprehend. Let us analyze these words further to provide a better definition and contextual meaning.

## Data: The First Rung of the Decision Making Pyramid

The word "data" in the past has meant the raw counts of things. In and of itself, data has no meaning. However, when data is aggregated, it can provide a path for providing contextual meaning. Here are some properties of the word, data.

- Data can be useful or not useful
- Data can be accurate or inaccurate
- Data can be valid or invalid
- Data can reliable or unreliable

Data is all based on collecting things, why we are collecting, and how we are collecting. If you look out the window and start counting the number of cars that are coming by, that would be known as data collecting. If you now note down how many cars pass through the stop sign, how fast each car is moving, or what the colors of each car is then these questions give contextual meaning to data collection.

Concepts such as usefulness, validity, accuracy, or reliability are all properties of data and become important, once the data is analyzed and for example, provided to larger groups of people. Data is usually as good as the measurement device that we use to collect it. It is also highly dependent on the person or automated device which is collecting it. There is a lot of time and effort spent on analyzing data errors, the formation of errors, and the errors caused by disseminating data. We use mathematics and the study of probability and statistics, to measure the error rates, and the accuracy or reliability of data.

However, in the context of today's information enterprise, data is at the bottom of the pyramid. First, because data is vast and plentiful and second, because data provides the information foundation upon which we make our collective decision making. A fuller and more current definition of data would be the following: data is a representation of facts, concepts, or instructions in a formalized manner suitable for communication, interpretation, or processing by human or by automatic means.

We have decided to separate data as its own information form, even though each of the mentioned information forms, such as text, images, and graphics, audio and video are all forms of data. The reason for this is because of the widely publicized digital revolution. The digital revolution has characterized data based on digits of 1 and 0. Before, in the analog equivalent, data had physical units or dimensions such as the power into your house is 110 watts, or the house temperature is set to 70 degrees Fahrenheit. However, based on a mathematical transformation from analog to digital conversion, analog data loses its dimensions and is transferred based on a digital representation of 1 and 0.

Data transfer and storage within a digital environment has resulted in ever greater amounts of data being collected, processed, and transferred. Based on the use of computers, data processing has become automated and is stored in digital formats. From the perspective of decision making, the large amount of data is both a boon and a curse. The vastly greater access to data has made the decision making process more complex and it has increased our collective need for greater information, knowledge and intelligence.

## Information: The Second Rung of the Decision Making Pyramid

So now that we have defined data, let's go on to ask – What's the difference in the two words, data and information?

Well, "information" is simply data that has contextual meaning. This is the reason why many people think that data and information is one and the same. But that is not true. Once we place contextual meaning to the data at hand, the properties of the data changes.

Data may be both useful and not useful, but information is mostly categorized as useful. This is because if we find that the data that we collected was useless, then we tend to dispose of it. Once we decide to keep data based on its contextual meaning, it becomes information. Hence, the characterization of information within the decision making pyramid as a rung is smaller than the bottom rung representing data.

Information can be characterized as collecting a lot of data together for one subject. Sets of data can be added to collections of related sets of data to form an information repository. Information is data that is comprehensible to human beings, that allows us to communicate and learn from others. In today's information enterprise, the amount of comprehensible information is also growing by leaps and bounds.

This may simply be because of the fact that we have a much more educated world, and people all around the world are communicating at a greater pace, along with creating and disseminating information. This is also because of the previously mentioned digital revolution and the advent of computers.

The World Wide Web is a great example of our collective information placed in easy access to all of us around the world. Every individual, business organization or government entity can, if they choose to do so, create information for the World Wide Web that is instantaneously available for all of the rest of us to glimpse and review. Over the years, libraries have served this purpose where everyone went to read books and share information. With the use of digital technology, this information library has now been virtually presented within the Web, so that it is available in electronic means.

However, living in today's information enterprise can also be characterized as "mind-boggling" since we are becoming inundated with more and more information. We all have access all the time to more information that we can possibly have time to put into context. We may not have time to decide what it means since it coming to us so quickly. As more digital information is shared between people and organizations, this poses a much more difficult challenge as to how to proceed for the future.

## Knowledge: The Third Rung in the Decision Making Pyramid

As we move up the Decision Making Pyramid, the next rung to consider is what we collectively call, knowledge. Knowledge is different from information in one crucial aspect. Knowledge results from the comprehension of information.

Knowledge is information with intrinsic value, because it has implications. In the case of the net-centric environment, the entire infrastructure is based on our collective knowledge. For example, if the next day's forecast shows that the weather is suitable for most military activities, and if enemy forces typically exploit good weather to launch an attack. Then based on this knowledge, the implication for a war fighter in Theater is that he should prepare to counter an enemy attack, possibly coming in the next day. This is the main characteristic of knowledge which is based upon information, experience, reasoning, and the situation at hand.

The body of knowledge is certainly smaller than the body of information. Hence, it is at a higher rung in the Decision Making Pyramid. As people spend time interpreting and understanding a body of information, they can create a body of knowledge that represents the collective thoughts of the group. This takes a lot of time and intellectual effort. While technology has greatly reduced the cost involved in assembling and storing data, and in transferring and storing information, technology has not done anything to make the process of creating knowledge any quicker or cheaper.

It is understood that creating knowledge takes an ability to comprehend, evoke thought and devote time for analysis and evaluation. This is especially true in today's information enterprise when there is so much more information available to wade through. People can become knowledge experts for a given subject, usually known as subject matter experts. In this information age, this means that they perpetually have to keep up with the pace of that subject. The general public then relies on these people to help bypass the costly process of wading through large bodies of information.

Since it is a difficult undertaking to create a body of knowledge from the greater pool of information, the credibility of knowledge experts becomes very important. Our society trusts people who possess educational degrees and certifications on critical skills. But for the general public, the level of knowledge possessed by experts is often hard to assess. You have to be able to trust the knowledge expert to give you an honest understanding of valid and reliable knowledge. Yet, you may lack the specific knowledge or intuition to know if the information and the ensuing knowledge are as reliable and credible as you need it. This then leads us to the higher rungs of the Decision Making Pyramid.

## Intelligence: The Fourth Rung of the Decision Making Pyramid

Intelligence is the next higher rung in the decision making pyramid, where we have to gather information to scintillate intelligence from disparate sources of knowledge bases. Sounds complicated, doesn't it? This is the reason why intelligence is precious and worth paying for.

Intelligence comes from the ability to synthesize various streams of knowledge, and even seemingly unrelated bodies of knowledge. Intelligence allows you the power to be able to make informed judgments about various ideas and propositions that may lie outside of your own direct area of expertise. However, just like the previous words of information and knowledge, intelligence needs to be gathered and put into one place. Intelligence for example, recognizes that certain patterns in nature repeat themselves, no matter where they may be found. A body of intelligence has enough wisdom to be able to spot such patterns and trends so that various bodies of knowledge can be put in context, combined and applied appropriately.

Intelligence as a thought process has been historically only ascribed to humans. However, with the advent of advanced computers, a new branch of computer science has evolved, known as Artificial Intelligence. However, computers are still bereft of common mechanisms for intelligence that humans take for granted. Intelligence usually requires a deep and perhaps intuitive understanding of human nature. This includes our five senses of sight, sound, smell, touch and taste. It also includes the human ambitions, our styles of intelligence, and our human motivations. This means that the possessor of intelligence is able to make judgments about representations of knowledge that lie outside of their own expertise. This comes back to our earlier problem with a body of knowledge. Intelligence can allow us to escape from the dilemma of the need to make judgments about knowledge experts who possess bodies of vast knowledge that you do not know about. Measuring intelligence means to be able to recognize your sharpness of judgment, along with measuring the amount of experience and perspective in a variety of knowledge.

In today's information enterprise, technology cannot confer intelligence, since intelligence takes a lot of time and knowledge to develop. It is quite true that people who possess advanced degrees may still lack intelligence, wisdom or judgment. This is why intelligence comes more and more at a higher premium. When you find good credible sources of intelligence, you hold on to it and try to grow it over time. Intelligence is often associated with wisdom which combines the seasoned experience of connecting and reviewing bodies of knowledge, together with a genuine grasp of human nature and the ways of the world to allow for the proper use of data, information and knowledge. Humans who possess wisdom cultivate their connections with other wise people or reliable knowledge experts, since this may be the most effective way to leverage and benefit from vast stores of knowledge in our information enterprise.

## Decision Making: The Highest Rung of the Decision Making Pyramid

Now that we have described the words, data, information, knowledge, and intelligence, we have deliberately placed the term, decision making at the highest level. This is to recognize that decision making should not be taken lightly. In most large organizations, decision making is a collaborative process. Informed decision making is the key element within the net-centric concept. Net Centricity transforms decision making abilities from central planners and a bureaucratic structure to edge resources and devices. Alberts, D.S., Hayes, R.E. (2005, April) talks specifically about this topic in their book to allow military commanders in the theater to be able to make intelligent decisions based on all information at hand.

Decision making sits at the top of the pyramid because its essential properties include each of the foundational items below it. Decision making, however, is actually necessary at each rung of the pyramid. When you are finished collecting data, you frequently have to make a decision as to if the amount of data that you have collected is enough. To make that judgment, the pyramid is stating that you need all of the other foundational items.

To figure out if the amount of data is enough, you have to find out if it is useful and does it have contextual meaning. If so, then this turns into a storehouse of information. Within this storehouse, you have to be able to categorize the information based on your particular subject matter and areas of knowledge. If you do not possess that knowledge, then you have to rely on external knowledge experts who will give you that information for you to comprehend. This then requires intelligence. You have to use your intelligence to find out if you possess the correct judgment and skill that allows you to make the final decision. So, the fact that even for a seemingly simple question: Is the amount of data that you have collected enough? You need to use the other rungs of the pyramid – data, information, knowledge, and intelligence.

As our information enterprise has become more vast and complex, the decision making process for using particular pieces of information has also become very difficult. The complexity arises from the plethora of information storehouses and knowledge bases that provide seemingly similar pieces of information. As part of the decision making process, you have to understand which pieces of information are relevant, what knowledge areas they do cover, and what pieces of intelligence it is allowing you to gather. Based on this process, it is important to then analyze a coherent decision making process that best uses your intellect and judgment.

This chapter provides a preamble to the net-centric transformation at hand. In most large organizations, it is important to recognize that a transformation process needs to take place to move from a command and control decision making process to a more matrixed net-centric decision making process. In an era of information overload, a command and control structure does not lend itself to analyze disparate sets of information and knowledge bases to come to a more informed judgment and decision making. Instead, a distributed and less hierarchical approach to decision making may allow an easier and more coherent decision making, and one that allows greater participation of personnel at more distant locations.

## INFORMATION SHARING STRATEGY

In the previous sections, we have discussed the structure of today's information enterprise, the various information forms, and necessity for a decision making pyramid. This section now addresses how information should be shared within the information enterprise to foster a net-centric environment.

As we create process and manage information, it is quite clear that information needs to be shared amongst our peers and to affected stakeholders in a coherent manner. In the case of our military, especially within the theater of warfare, how, when and where we share information becomes one of the most crucial of objectives, since the consequence is a matter of life and death. This requires taking deliberate information sharing measures that include the processes of information collection, analysis, and integration. An informed decision making process can be convened when each of these processes have been fully evaluated. This information sharing decision making builds an increased situational awareness and allows continued efforts in improving and managing knowledge.

As described within the DoD CIO (2007) DoD Information Strategy document, our defense strategy promotes an universal Information Sharing Value Chain that can be used by any set of stakeholders regardless of the type of military or civilian organization. This value chain can be used regardless of the organization's mission domain, community or the organization's set of unique processes for managing their own information. The overall goal is to discover and collect information so that we continuously add value at each stage of the process. This allows us to become best informed as a decision maker.

Figure 2 provides an illustration of the Information Sharing Value Chain as excerpted from DoD CIO (2007) and can be described as the "Discovery to Decision Making" continuum. This continuum provides a perspective of continual information collaboration and enhancement of data to create relevant knowledge and actionable intelligence.

Let us now review each of the steps of this value chain, and understand the stepwise actions based on an ongoing mission need.

- **Mission Need** – Every mission within the military, government, and civilian organizations have an ongoing set of objectives that define the mission and information requirements to manage this mission.
- **Discover/Collect Information** – Based on the mission need, it is important to discover and collect information so that the organizations can function and operate on their own mission needs. This discovery entails an ongoing sensing of what is occurring and collecting the necessary and available data based on the use of various sources, tools, systems and relationships.
- **Process** – The process of analysis and evaluation involves finding contextual meaning of the data that is gathered, assimilating useful data that is captured and stored as information, determining

*Figure 2. Information sharing value chain*

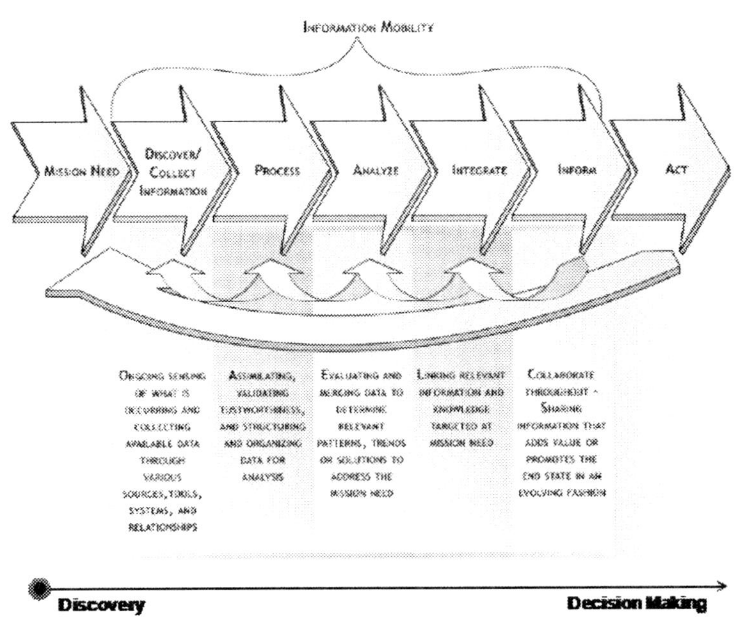

the trustworthiness of the information, and structuring and organizing the data for analyses to form information and knowledge

- **Analyze** – This analysis process is part of the mission need to fulfill the organization's objectives to evaluate and merge useful data and information to determine relevant patterns, trends or solutions. This analysis process is part of the higher level of evaluation that transforms data into information, and information into knowledge.
- **Integrate** – This next step for integration recognizes the need for organizations to work together with other affected members and stakeholders to link relevant information and knowledge that is targeted for the mission need. This integration step is a key step in gathering intelligence, since it requires the ability to glean information and knowledge from disparate sets of storehouses.
- **Inform** – Once the knowledge and intelligence has been integrated, it is time to collaborate and share this information with others and disseminate the appropriate information to trusted partners. This collaboration adds value to the information since it promotes the end state of the organization's mission needs as the information enterprise continually evolves.
- **Act** - Now that you have been adequately and summarily been informed of the organization's needs, it is time to act. This then involves the actual decision making process of how to act, what to do, and how to move forward. In the case of DoD and the military, this decision making process may also lead to a large implementation and deployment process that requires decisions to be made at each step of a deployment.

It is important in today's information enterprise for all organizations to have an information sharing strategy. This leads to an informed and a more effective decision making process that helps the organization as a whole, and promotes a more organization communication mechanism

## TOMORROW'S ENTERPRISE: FUTURE TRENDS

This last section of the chapter looks to the upcoming future and provides some thoughts as to changes that are coming to today's information enterprise environment. Since there are so many opportunities and technology challenges that are part of the information industry, we will only focus on a few of the large paradigm shifts within our environment.

### Next Generation Internet

We all recognize that the Internet is rapidly transforming business, markets, and customers. Within the civilian and market driven economy, every industry from financial services to health care, to the electronics and communications industry to education initiatives are being rapidly transformed. The Internet and the World Wide Web have become a global library and a distributed storehouse of information. The Web is also being used to provide products and services are changing the ensuing supply chains in every market throughout the globe. The Web is also allowing the convergence of computers, networks, storage and wireless technologies to create both opportunities and threats.

The Next Generation Internet uses IP Version 6 as opposed to the current transport protocol that is based on IP Version 4. For the current Internet infrastructure, a complete switch to the new IPv6 protocol brings a new transformative change to today's information enterprise. The Next Generation Internet plans to merge telephony and video into a new interactive and sensory experience that applies to all industries including entertainment, retail, healthcare, education, and science and technology markets. For example, it is possible to move ahead on development of real-time video with interactive multimedia that can be embedded within clothing or military uniforms. It can also allow cars to navigate and drive by themselves and provide an increase in virtual reality and computer simulations. The Next Generation Internet plans to become more sensory in nature that is aware of its surroundings, becomes adaptive and develops a digital personality. This new Internet will also be available to more people around the world, and can iteratively contain the collective intelligence of a vast global population.

## Real-Time Collaboration

Based on technology changes, it is becoming more prevalent for people in distant locations to communicate among their peers through real-time collaboration. This can be a 24 by 7 activity, where interactive video allows multiple groups to collaborate on a common thought process for a common mission. A rich media experience involving multiple individuals that is interactive and multi-sensory and can allow different worlds to come together on a continuous basis.

This collaboration also adds the ability for individuals to search and find what they are looking for based on semantic means. Customers can define what they want, and vendors can define what intend to sell to the right set of customers. This buy/sell process can actually be transformed based on enhanced collaboration, which adapts to inputs from various members and organizations real-time. Other collaboration tools such as the creation of Personal Portals can allow each member to his or her own digital needs and desires that can live with them within their cars, homes or portable devices. This portal can then be shared with external members in distant locations so that the information enterprise becomes more aware and can act in an expedient manner accordingly.

Collaboration through the use of multiple computing platforms has been successfully accomplished through the use of enterprise portal software. As technology proceeds further, enterprise portals will be used more and more as a framework for integrating information, people and processes across organizational boundaries. Collaboration based on an enterprise portal offers a secure unified access point, often through a web-based user interface, and is designed to aggregate and personalize information through application-specific portals. Enterprise portals allow content management and content distribution to be de-centralized, so that information updates can occur automatically across multiple organizations.

## Convergence of Electronics and Transportation

As networks across the globe adopt wireless mechanisms, it is becoming possible to create cars and personal vehicles that are completely functional based on the Internet or a secure Intranet. A wireless secure intranet car can be part of a critical transportation and infrastructure link within today's cities and country side. This means that cars can effectively navigate by themselves based on GPS input and transportation network ongoing messaging. All of the points of interest, including retail stores, restaurants, government buildings, and a person's place of work can be part of the common grid that sends continuous messages to a vehicle as it is moving.

This convergence of electronics and transportation has been happening for a long time. However, the pace of change is about to significantly increase based on the migration to a Next Generation Internet, and the establishment of a broadband network infrastructure that includes wireless signals using broadband transmission. The concept of a wireless internet-ready car is coming as another critical link in the wireless mobile network that is being constructed across the country.

This type of technology would have voice-recognition systems built into the dashboard that automatically finds the restaurant, conducts purchases, or locates a destination for a trip. GPS satellites are already linked to navigation devices that offer location-based services for everyone. This may also allow auto companies to create service portals that helps the consumer and generates additional revenues for the manufacturer.

## Network and Communications

The next generation net-centric capabilities may mostly be in developing improved network and communications across the globe. This includes the use of communications equipment, data, and software to enable the networking of people, along with automated sensors, and different computing platforms. Communications technology uses a host of transmission methods that includes use of radio transmission, microwave, infrared signals, or laser beams. Communications lines offer the ability to send massive amounts of data through larger network segments and offer it for global distribution. The communications can be via fiber optic cables, microwave towers, low altitude or high-altitude satellites. Network and communications design must allow rapid communications between individuals in all type of services, along with rapid sharing of data and all types of computing platforms.

For the past number of decades, the DoD has been using the NIPRNet (Non-Classified Internet Protocol Router Network) for most of its information, communication and transmission. For information that requires a higher classification, the SIPRNet (Secret Internet Protocol Router Network) provides a secret information network. This network architecture needs to be upgraded to the latest communications technology so that tomorrow's military can take full advantage of broadband data.

Satellites and satellite communications have been crucial for enabling remote communications across the globe. Satellite technology has been used for providing imagery, navigation of vehicles, displaying weather information, or providing early warning capability for a ground based missiles. At this time, the U.S. maintains a number of orbital constellations that provide Intelligence, Surveillance, and Reconnaissance (ISR) capabilities. This includes for early warning, imagery, and signals intelligence. The expansion of satellite technology is crucial in greater information sharing of multimedia data.

## Unmanned Vehicles

The military is starting to use a number of unmanned vehicles (UVs) for conducting crucial operations for intelligence, surveillance and reconnaissance. These UVs can consist of Unmanned Aerial Vehicles (UAVs), Unmanned Ground Vehicles (UGVs), or Unmanned Underwater Vehicles (UUVs). UVs can provide surveillance data based on use of broadband transmission and communications. For net-centric warfare missions, the use of UVs are preferable since they allow a large amount of intelligence data to be known by military forces beforehand, and may lessen the necessity for combat. UVs can serve as operational nodes in theater to relay messages across a net-centric warfare network. All of the Armed Forces, including the Army, Navy, and Air Force are ramping up their deployment of UVs based on the rapid change in multimedia technology to provide up to the minute information of tomorrow's battlefield.

## Nanotechnology

The use of nanotechnology as a new branch of science allows research and development in new materials and rapid changes in battlefield equipment. Through the use of nanotechnology, it is possible to make weapons smaller and lighter, and build and deploy miniaturized network sensors that may detect, locate, identify and track potential threats within the theater and battlefield.

Research organizations are currently conducting research in various aspects of nanotechnology that includes for example, creating a heat-resistant coating for nose-cones and propeller shafts for airplanes and ships, or to prepare combustible additives for changing the performance of a rocket propellant. The ability to create miniature devices helps to create for example, a handheld device for detecting chemical or biological weapons, or to create a flexible exo-skeleton skin for a soldier's protective gear that makes the gear many pounds lighter, yet it makes the equipment bullet-proof. Research and development in nanotechnology will probably proceed for many decades, and the full breadth of new technologies has yet to be fully fathomed.

## REVIEW OF CHAPTER GOALS

The goals of this chapter were:

1. To address a glimpse of today's information enterprise:
   - Address the complexities and how the information industry is rapidly changing.
2. To provide an understanding of how an information enterprise is structured:
   - Discuss the different types of information in business and their core competencies.
3. To address the information forms and functions within the enterprise:
   - Provide definitions of text, images, voice, audio/video and data
4. To provide an original concept that defines a decision making pyramid:
   - How data, information, knowledge, and intelligence in different rungs help in the decision making process.
5. To have an understanding of an information sharing strategy for a large enterprise:
   - Provide the steps of an information sharing value chain
6. Last but not the least, to discuss a number of technologies that are upcoming as future trends:
   - This includes next generation Internet, real-time collaboration, electronics and transportation, network and communications, unmanned vehicles, and nanotechnology.

## REFERENCES

Alberts, D. S., & Hayes, R. E. (2005, April). *Power to the Edge*. Command and Control Research Program Publication Series. Available from www.dodccrp.org

Ghosh, S. (2006, July). *Electronic Records Archive (ERA) Technology Evolution Plan*. National Archives and Records Administration, Lockheed Martin, 2nd ERA Technology Evolution Council Meeting.

Ghosh, S., Miller, C., Przysucha, J., & Tiemann, M. (2005, January). *Advancing Enterprise Architecture Maturity, Version 2.0*. Industry Advisory Council White Paper.

Office of the Assistant Secretary of Defense for Networks and Information Integration, Department of Defense Chief Information Officer. (2003). *DoD Net-Centric Data Strategy* (Version 1.0).

Office of the Assistant Secretary of Defense for Networks and Information Integration, Department of Defense Chief Information Officer. (2007) *Department of Defense Information Sharing Strategy* (Version 1.0).

Spewak, S. H. (2002). *Enterprise Architecture Planning: Developing a Blueprint for Data, Applications, and Technology*, (2nd Ed.). Chichester, UK: John Wiley & Sons.

The International Engineering Consortium. (2001). *Web ProForm Tutorials*. The New Information Industry. Retrieved from www.iec.org

# Chapter 3
# Measures of Interoperability

## CHAPTER CONTENT

As you explore Chapter 3, it will cover the following topics:

- A Broad Definition of Interoperability
- Types of Interoperability
- Interoperability based on Loose Coupling
- Measures of Interoperability
- Architecture Strategies for Greater Interoperability
- Interoperability in Large-Scale Distributed Systems

## CHAPTER FOCUS

Now that we have discussed today's information enterprise, this chapter focuses on the larger concept of interoperability. The definition of this word is not simple and a lot has written about this topic, so

DOI: 10.4018/978-1-60566-854-3.ch003

this chapter focuses on providing an objective definition of the subject matter. It then defines particular types of interoperability and how these types are measured. It defines the concept of loosely coupled systems, and how to obtain greater interoperability through looser coupling. It describes an objective way to measure interoperability based on the LISI profile in use by the DoD and NATO. It then provides an understanding of architecture strategies for achieving greater interoperability. It ends the chapter by discussing interoperability measures within large-scale distributed systems.

## A BROAD DEFINITION OF INTEROPERABILITY

This chapter explores the other important concept that is within the title of this book, namely the word, Interoperability. As a concept, Interoperability goes hand in hand with the Net Centric vision. Since interoperability is a complex topic, this chapter provides a discussion on its definition and how interoperability should be measured across the information enterprise.

A popular perception within the technology industry is that interoperability is synonymous with connectivity. However, true interoperability is much more than just connectivity. Making two entities interoperable involves an integration of operational concepts and scenarios, policies, processes, and procedures. Within the net-centric context, interoperability is considered to be a desired but unattainable goal rather than a condition that can be quantified. It is understood that systems may choose to be either interoperable or not interoperable based on their function and day-to-day operations. For this reason, developing and applying precise measurements for this area is multidimensional and complex. The DoD admits that within their military forces and equipment, serious interoperability deficiencies exist today. For years, each of the Armed Force Services that includes the Army, Navy, Air Force and Marines have all had their own set of applications and platforms that were by design made to not be interoperable with other systems. One of the detractions of a command and control structure is the issue of low interoperability – this has been identified in all recent, allied, joint, and combined operations and exercises.

Hence, most government agencies along with large corporate organizations recognize that there is a greater need for interoperability. This notion is also shared across the globe by many other countries along with the respective country's military. Interoperability has been identified as a major issue by NATO which has to combine foreign forces along with disparate systems from a host of different countries, all following a separate set of information technology standards.

### Defining Interoperability

The following definitions provide a broad definition of Interoperability based on different DoD directives and instructions.

**Interoperability Definition based on the DoD Joint Publication:** (Joint Pub 1-02, 2001)

- The ability of systems, units, or forces to provide services to and accept services from other systems, units, or forces, and to use the services so exchanged to enable them to operate effectively together

**Interoperability Definition based on Chairman of the Joint Chief of Staff Instruction:** (3710.01b, 2001)

- The condition achieved among communications-electronics systems or items of communications-electronics systems equipment when information or services can be exchanged directly and satisfactorily between them and/or their users.
- The degree of interoperability should be defined when referring to specific cases. For the purposes of this instruction, the degree of interoperability will be determined by the accomplishment of the proposed Information Exchange Requirement (IER) fields

**Interoperability Definition based on DoD CIO Global Information Grid Capstone Requirements Document:** (GIG CRD, 2001)

- Ability of information systems to communicate with each other and exchange information
- Conditions, achieved in varying levels, when information systems and/or their components can exchange information directly and satisfactorily among them
- The ability to operate software and exchange information in a heterogeneous network (i.e., one large network made up of several different local area networks)
- Systems or programs capable of exchanging information and operating together effectively

The reason to offer multiple definitions is that there probably cannot be a precise definition of interoperability that incorporates all of the measures and features that everyone would like. The definitions may also change due to differing expectations that are constantly changing. Based on net-centric capability definitions, a new set of interactions between systems may need to be detailed.

For the purposes of this chapter, the best line to remember may be that interoperability is more than just information exchange. It includes systems, processes, procedures, organizations, and missions over a sustained life cycle and must be balanced with information assurance features.

We now go ahead and address particular types of interoperability, as well as ways to properly measures interoperability between and among systems. However, it is recognized that there is no method today for tracking interoperability in a quantitative manner, and on a comprehensive or systematic basis.

## TYPES OF INTEROPERABILITY

Since interoperability has many facets that include interoperability within systems, data, processes or organizations, we further discuss the different interoperability types. The definition of interoperability based on the references above, encompasses both a technical and an operational capability type. The following definitions have been arrived at, based on information provided within the GIG Capstone Requirements Document (GIG CRD, 2001).

- **Technical Interoperability** - refers to the ability of systems, units, or forces to provide services to and accept services from other systems, units, or forces. This addresses issues of technical connectivity among systems, sharing based on data and file exchange, common networking schemes, and other communication related scenarios.

- **Operational Interoperability** - refers specifically to the ability of systems, units, or forces to use the services that are exchanged to enable them to operate effectively together. This type of interoperability measure addresses the degree to which value is derived from the technical interoperability capability.

Technical interoperability from a systems perspective can be further defined based on a set of system conditions that is achieved among communications-electronics systems or communications-electronics equipment. The degree of technical interoperability can be measured based on data or information exchanged directly between systems to the satisfaction of the end users. System test cases can be created that defines the degree of interoperability for specific test cases and exchange specifications.

For two command and control systems to effectively interoperate, they must be able to exchange relevant bit-streams as well as to interpret the data bits they exchange based on a consistent set of data definitions. This means that technical interoperability places demands at multiple levels of system development and deployment that ranges from physical interconnection to correct interpretation of software applications based on data that is provided by other applications.

Degrees of technical interoperability can be measured within the following:

- Automated sensors that generate measurement data
- Network and communication channels that transmit data
- Computing platforms that process application data
- Weapon systems that direct messages based on input data

Operational interoperability goes beyond the technical interoperability measures to include people and procedures that interact based on an end-to-end process. It is more difficult to measure degrees of operational interoperability. Since these measures need to include a wider view of systems development life cycle that includes configuration management, software versioning, and unit, system and integration testing, therefore, implementation of operational interoperability procedures need to be included within the system development and deployment life cycle.

Other types of interoperability include data interoperability, syntactic interoperability, and semantic interoperability. Within the DoD, data interoperability issues are usually discussed at length between system managers before systems are modified significantly. A set of approaches to reach greater data interoperability includes:

- **Use of a Single Data Definition for all systems** – Program management may decide that use of a single set of definitions is mandatory for all applications. This means that definitions used locally are no longer optimal. This decision avoids the time most product teams take to agreeing on a common set of definitions.
- **Object Orientation** – Currently, most organizations are using data in an object-oriented fashion that allows data properties to be defined as public and private and allows the black-box approach to segmenting data. Systems can then encapsulate the internal details of the data and share information as it deems necessary.
- **Use of an Extensible Data Model** - This approach uses an extensible data model that standardizes the way system data is scaled as the system grows and integrates with multiple systems.
- **Adoption of Extensible Markup Language (XML)** – Most DoD and federal organizations have adopted the approach to use of XML for their data transfer based on establishing agreement based

on contents and meaning of the XML schema. This allows system and application data communicated based on XML to conform to a common schema. This is similar to mandating a single data definition for multiple organizations.

Other Interoperability types include:

- **Syntactic Interoperability** - requires the use of a common syntax as offered by the use of XML or SQL standards. Syntactic interoperability is usually a pre-requisite to semantic interoperability. Syntactic involves the use of a common data format and common protocol structure. The idea is to resolve ambiguities in the data structure so that there is no miscommunication. From a transactional perspective, this would include resending any message that appears to be garbled or incomplete.
- **Semantic Interoperability** – This rides on top of syntactic interoperability. The expectation is that if two systems are semantically interoperable, then when they exchange information, the meaning of that information is automatically interpreted by the receiving system accurately enough to produce useful results. This is a much more difficult environment to achieve and it is dependent upon the active cooperation of the end users of both systems.

## INTEROPERABILITY BASED ON LOOSE COUPLING

Achieving interoperability may be both a blessing and a curse, and the answer depends largely based on a careful analysis of the total set of users and stakeholders. In general, having greater interoperability between two systems provides a benefit to all system users. However, if a system attack occurs by hackers and subversive elements, a high degree of interoperability between critical systems may be a curse, since it could result in a total shutdown of large computing environments.

From an information assurance perspective, it is also important to ensure that private and classified information remain secure, and greater degrees of interoperability may work against that goal. This has led to system architecture models that allow system owners to hide their private details within the system boundary, but share information with customers and vendors based on a selective process of exposing data elements when necessary. This type of an architecture model follows a loosely coupled integration concept.

The ideal state would be for systems to have a high degree of external interoperability, but the individual system details are kept private and are only known to the actual system owner. This may be achieved based on systems which have a greater degree of interoperability but employ a loosely coupled approach to their systems integration. Loose coupling describes an approach where integration interfaces for systems are developed with minimal assumptions between the sending/receiving parties. This reduces the risk that a change in one application or module within an integrated system will force a change in another application or module.

The concept of loose coupling may have multiple dimensions. This means that integration between two applications may be loosely coupled in time, or loosely coupled based on the use of different data formats. When we discuss the topic of Service Oriented Architecture (SOA), you will find that the loosely-coupled model is a central theme within the concepts of SOA. The idea for example, is to use

Web Services so that the coding implementation of one system is hidden from the consumer of that Web Service.

In the past few decades, most computing platforms were known for being very closely coupled. Tight or close coupling means that when there two systems are simultaneously operating, they allow direct inter-system communications between them during program execution. Within a distributed environment, data communications standards such as Microsoft's DCOM (Distributed Component Object Model) or CORBA (Common Object Request Broker Architecture) employ close coupling methods, since they allow remote services to occur, but allow users to keep control of their local machines. For example, Microsoft's framework allows a single script to invoke commands from both systems.

The wide use of the Internet and the World Wide Web has shifted the distributed computing paradigm from a traditional closely coupled method to a loosely coupled one. The World Wide Web is intrinsically loosely coupled. It provides a loosely coupled network platform that allows developers to change the implementation at either side of the connection.

## Measuring Coupling

The ability to measure coupling between systems greatly affects the system interoperability. The degree to which systems are coupled can be noted by the number of changes in data elements that occurs in the sending or receiving systems and determining if the automated systems would still continue to communicate correctly. Coupling can be measured by tracking the changes to data elements:

- New data elements being added to messages
- Data element order being changed
- Names or structures of data elements being changed
- Data elements being omitted

## Methods for Decreasing Coupling

As we stated earlier, it is best to lower the level of coupling between systems, since it makes it easier to achieve a higher level of interoperability quickly. The assumption here is that systems are still distinct from one another and one does not require being a subset of another.

System interface can be more loosely coupled, if publishers of data transmit their messages based on a flexible file format such as XML. This enables subscribers to publish clear definitions of how they would subsequently use this data. For example, a subscriber can publish a collection of statements used to extract information from a publisher's messages by sharing the relevant XPath expressions used for data transformation. This allows a data publisher to test whether their subscriber's extraction methods will fail if the published format changes.

Another mechanism to promote loose coupling is to enable data replication to occur across different systems. This makes sure that there is data interoperability between the systems, but hides the actual system functions of a particular system. Message-Oriented Middleware may allow systems to be decoupled based on time, since information can be sent and received asynchronously. However, these mechanisms may also have ill effects that may raise the level of synchronization that is necessary between systems.

## System Parameters for Greater Interoperability

Based on the premise that we want both a greater degree of interoperability and measure of loose coupling, the following provides a set of system parameters that all parties can agree with, as to the future state:

- **Scalability** - the system must have an architecture that is fundamentally scalable
- **Adaptability** – ensure that network input and output, along with bottlenecks are able to deal with caching, job batching, and allows for other systems to piggyback on top of each other
- **Reliability** – make sure that the system has good redundancy, replication, and recovery techniques and is able to detect and mask failures
- **Simplicity** – it is best to reduce complexity wherever possible, and limit the overhead necessary to make the system a manageable commodity
- **Heterogeneity** – ensure that the system environment supports for a heterogeneity in terms of programming tools, front end user interface, and databases
- **Security** – raise the criticality of information and call for protecting the integrity and privacy of information
- **Real-time support** – use existing and future multimedia applications to support the need for real-time traffic control for interactivity and quality-of-service requirements.
- **Load balancing** – allows the system to load balance its data processing and storage and allow for both static and dynamic adaptive algorithms.
- **Ease of maintenance and implementation** – make sure that the system is able to be maintained and administered based on low cost and that the system provides a commercially viable solution

## MEASURES OF INTEROPERABILITY

Large organizations such as the DoD and NATO have been working on codifying interoperability measures for a while. Within the DoD, interoperability is assessed based on meeting a number of criteria. These criteria include, for example, a focus on technical standards, the goal for creating a Common Operating Environment (COE) for military systems, an enterprise level data model structure, and system level certification criteria. The idea is to figure out how individual system compare against these set of criteria or standards.

It is clear that precise measurements for measuring interoperability is difficult, and may not truly serve the greater need of the large organizations. Instead, a global scale that judges the interoperability levels of most systems is easy to understand and proliferate. A proper measure of interoperability needs to include all of the system operational concepts and scenarios, policies, processes, and procedures. Interoperability measures can be characterized in the following manner:

- Technical compliance measures
- Systems interoperability measures
- Operational interoperability measures
- Organizational and cultural measures

*Figure 1. LISI Profile Interoperability Levels*

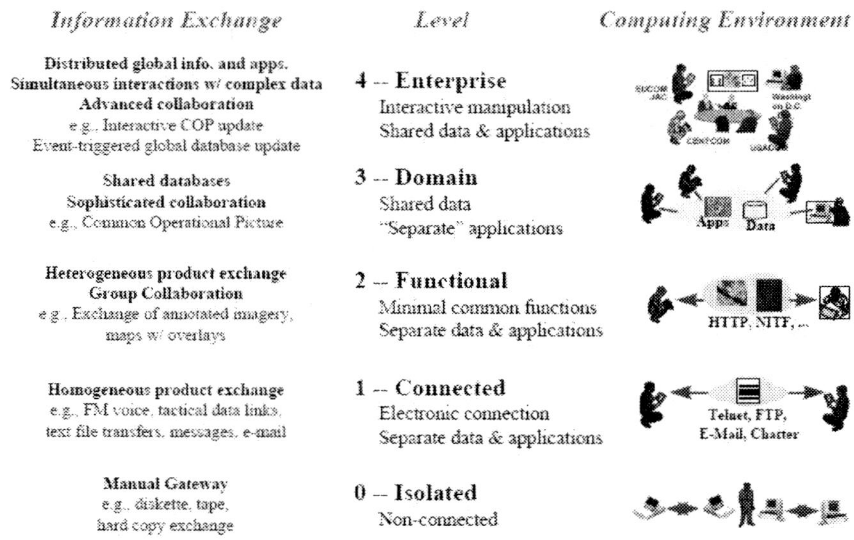


Based on this guidance, the most well-known scale for measuring interoperability is the LISI profile, where LISI stands for Levels of Information Systems Interoperability.

## LISI Profile

The LISI profile was initiated by the DoD in 1993, originally as part of the DoD C4ISR Integration Task Force, and provided within the document they created (DoD C4ISR Working Group, 1998). However, as a model to follow for interoperability, the LISI profile is still relevant now and for the upcoming future. LISI has been developed as a reference model that provides a process for assessing the interoperability of an information system. The process includes defining, assessing, and certifying the degree of interoperability that is achieved between organizations or systems.

The LISI profile has been divided into five categories as has been shown in Figure 1:

- **Level 0: Isolated (Manual Gateway)** – This level signifies that the system is essentially non-connected or stand-alone – it uses of manual mechanisms to transfer data based on removable disks
- **Level 1: Connected (Simple Electronic Exchange)** – This level means that there is an existing electronic connection, separate data and applications
- **Level 2: Functional (Complex Media Exchange)** – This level allows common functions between systems to be shared, but systems contain separate data and applications
- **Level 3: Domain (Shared Application & Databases)** – This level means that two systems share a common data source, but have separate front-end application for their users
- **Level 4: Enterprise (Cross Domain Interactive Manipulation)** – This is top-most level and offers multiple user manipulation of a common set of applications and data sources; it offers shared data and shared applications

*Figure 2. LISI PAID Profile Attributes*

| Nature of Operational Information Interaction | Corresponding Interoperability Level | | Implications | | | |
|---|---|---|---|---|---|---|
| | | | P | A | I | D |
| Cross-Domain Interactive Manipulation | Enterprise | 4 | Enterprise Level | Interactive | Multiple Topologies | Enterprise Model |
| Shared Applications & Databases | Domain | 3 | Domain Level | Groupware | World Wide Networks | Domain Model |
| Complex Media Exchange | Functional | 2 | Program Level | Desktop Automation | Local Networks | Program Model |
| Simple Electronic Exchange | Connected | 1 | Local/Site Level | Standard System Drivers | Simple Connection | Local |
| Manual Gateway | Isolated | 0 | Access Control | N/A | Independent | Private |

The following diagram is excerpted from a DoD C4ISR Framework document and provides an understanding of each of the LISI levels.

The LISI profile within each level is bound by four enabling attributes of interoperability. These attributes are collectively referred to as PAID by the LISI profile, where PAID refers to:

- Procedures
- Applications
- Infrastructure (hardware, communications, security, and system services)
- Data

This hierarchy is then reflected in Figure 2 as a LISI PAID Profile.

The definitions for each attribute are the following:

- **Procedures** – reflects the doctrine, policies and procedures, architecture, and technical standards that enable systems to exchange information.
- **Applications** - reflects the software applications that enable the exchange, processing, and manipulation of data for the user
- **Infrastructure** - reflects the environment (hardware, networks, systems services, etc.) that enable the system interaction
- **Data** - reflects the formats and protocols that enable data interchange, along with the shared semantics that enable information interchange

Based on the diagram above, each of the PAID attributes follows a stepwise implication when you go up the LISI scale. For example, in the case of procedures, it addresses access control procedures at the stand-alone level, local or site procedures at the connected level, organizational program activities at the functional level, domain procedures for the shared applications and databases, and enterprise-level procedures at the cross-domain level.

It is important to investigate the LISI profile further since it provides a fairly detailed model of interoperability as well as a mapping between the model and implementation technologies. LISI assesses the level of interoperability attained between systems and not between users. Once system-to-system interoperability issues are isolated, the ability to address user interoperability issues is vastly improved. This allows system owners and program managers to address interoperability issues based on functional training needs and shortfalls, differing operational methods and procedures, and difficulties in user-to-computer interactions.

However, the DoD recognizes that the approach taken by LISI can become old and dated as new technology emerges. In this case, system owners need to be able to use a stoplight reference model like LISI as a guide, but be able to adjust as necessary to a new technology environment. You should also realize that a periodic set of interoperability assessments is needed throughout a system's life cycle and when the system changes in a major manner.

## ARCHITECTURE STRATEGIES FOR GREATER INTEROPERABILITY

This book has been preaching the sentiment that Net Centricity and Interoperability go together. The goal for future net-centric systems is to gain a higher level of interoperability in the battlefield and in the garrison back home. The command and control systems that have been in use for the past few decades have been characterized by a low degree of interoperability.

External integration of command and control systems has had the following set of detractions:

- System exchanges are complex and use proprietary, custom interfaces
- System parameters change frequently and are difficult to predict
- System may be operational at multiple organizational levels, this may include interagency, across DoD Services or joint forces

So to achieve and maintain interoperability, it is critical that a broad and yet detailed architecture vision is established that clearly denotes the objective relationship among systems and organizations. DoD directives and instructions have, over the years, mandated that new systems are able to operate within a joint and integrated architecture. The DoD Architecture Framework (DoDAF) provides a framework for formulating the overall architecture of an organization. This architectural framework will be discussed in more detail in later chapters. However, it is important to note that acquisition of new systems is heavily dependent on maintenance of clearly defined enterprise architecture.

A clear definition of the architecture of an information system is that it provides the structure or structures of the system. This includes the system components, the externally visible properties of these components, and the relationships among them. An architectural perspective or framework helps to organize the system complexity and the interoperability challenges in ways leads to more coherent relationships.

Architecture strategies provide a pictorial description for the design of a system and denote how it will be conceived, developed, and operated. System architecture provides the underlying blueprint for a more detailed design and implementation about the components of the system.

With regard to achieving greater interoperability, there are a number of architectural characteristics that may be considered as appropriate quality attributes that can be measured. The architecture strategies that are worth exploring for greater interoperability are the following:

- Properly defining interfaces and layers,
- Use of operational and technical standards
- Foster greater data interoperability

## Interfaces and Layers

Interfaces usually arise when systems want to connect and exchange data and information amongst each other. Systems that perform a variety of functions are normally composed of multiple subsystems or components. These subsystems or components may either interact internally within the information system or connect to external systems. In this case, architecture helps to design and partition a system so that it interoperates with existing and future systems.

Layers facilitate making command and control or net-centric systems interoperable when faced with the presence of rapidly changing technologies and multiple technical mechanisms. An architecture strategy such as layering, allows one to design a system of systems that can adhere to technology standards, promote scalability, provide decentralized operation, support data security and provide flexibility.

Since systems are modular in nature, a modular decomposition of a system is both horizontal and vertical. Vertical decomposition usually refers to interfaces between discrete systems that maybe within the same infrastructure layer. For example, in the messaging layer, a standard message format may be used by different applications to exchange information. Horizontal decomposition of functions may promote layering in another manner. For example, the separations of bit transport technologies, transport protocol, and applications.

Middleware provides a great example of the layering principle since it separates the applications from the operating system, or the higher level applications. Middleware services are at a different layer, which are sets of distributed software between the application and the operating system. The middleware layer also interacts with the network services on a system node in the network. By decreasing the dependence of applications on a particular operating system, middleware increases the ease of moving applications to new technology over time.

Best practices with regard to architecting interfaces and layers may be the following:

- **Defined Interfaces** - A well-designed and documented interface permits organization program managers to develop systems that can be divided into more manageable pieces. This then results in faster system development since developing the interfaces can happen in parallel and then be brought back for complete systems integration
- **Encapsulation** – The strategy is to encapsulate the internal details of a system component which may change over time. This permits modular changes of the system based on versions and various deployment technology. An encapsulated interface allows changes in internal implementation of portions of a system to be transparent to other external systems.

- **Reduced Interaction** – The goal should be to reduce the complexity of inter-system dependencies. This can then facilitate more rapid reconfiguration of systems to meet operational requirements.

## Use of Standards

An essential aspect of architecture strategy is to establish standards in the way organizations behave with regard to technology. In general, standards define common elements for information systems. These standards are deemed to be technical in nature and include aspects of user interfaces, system interfaces, representations of data, protocols for the exchange of data, and interfaces accessing data or system functions.

With regard to achieving greater interoperability, the following set of strategies needs to be followed with regard to adhering to standards.

- **Data Communications and Technical Interface Standards** - These standards exist at the physical and data layers of a system or organization. This includes the interfaces among the data systems, transmitter/receiver, and the handshake. This ensures that the systems are mutually compatible with regard to signals that are automatically exchanged between them. The signals could be for example, be waveforms or modulation techniques.
- **Messaging Standards** – There are a number of aspects of messaging standards between connected systems. These communication aspects are the data elements, or the types of information to be transmitted, the data items that transmit allowable values of that information, and the message format which provides the order in which the data is arranged.
- **Database and Application Standards** – There are a large number of database and software application standards that need to be negotiated for systems between organizations. This includes the exact set of data fields and variables, the user front-end look and feel design standards, and how the data is stored and interpreted.
- **Operating procedures** – Even though operating procedures may not constitute as standards for many systems or organizations, they are a key artifact in ensuring that standards are accepted and followed. Operating procedures are usually associated with day to day operations of multiple systems. They refer to procedures that need to be followed by data system operators, for example, interface procedures for the establishment of data links and exchange of tactical data.

The use of commonly-used standards greatly increases the chances for system interoperability. With regard to government and civilian contractors, standards are important because they are accepted by multiple vendors, and they increase the likelihood that a collection of systems from diverse sources will be able to interoperate.

However, it has become generally accepted that although standards are certainly beneficial, simple adherence to standards is not sufficient to guarantee interoperability. Interoperability measures need to be instilled within the system processes, so that accepted standards and compliant products are chosen. Then the systems life cycle process needs to work with accepted standards for different releases and versions of the product.

## INTEROPERABILITY IN LARGE SCALE DISTRIBUTED SYSTEMS

In this section, we discuss the challenges of ensuring interoperability among large scale distributed systems. In the past few years there has been an explosion in the amount and diversity of information available across networks. For example, the proliferation of Internet, Intranets and the ongoing advances in the World Wide Web, have fueled the development of a wide range of data-intensive applications and information dissemination systems. These distributed systems are helping to deliver information contents to users at home and on the road.

At this time, most organizations and enterprise computing systems are trying to figure out how to provide uniform and scalable access to information. Our information enterprise usually consists of multiple disparate information sources and repositories including:

- Databases
- Object Stores
- Knowledge bases
- File systems
- Digital Libraries
- Information Retrieval Systems

It is widely recognized that information sources change constantly and users are faced with the daunting challenge of navigating, collecting, evaluating, and processing data in this dynamic and open information universe. As the amount of information mounts, decision makers who often need information from multiple information sources, are unable to get and fuse information properly. The information sources may not provide information in a timely fashion, due to the unpredictable state of networks and because of the heterogeneous and evolving nature of the information sources.

### Next Steps Toward Greater Interoperability

The goal for tomorrow's distributed information system is to be open and interoperable, rather than be a static data delivery system. This is particularly important for the future net-centric capabilities. Two critical information requirements of such a system are the following:

- Support of an extensible distributed object management architecture
- Provide dynamic interoperability among diverse information sources and between the information consumers and information producers

With regard to the use of object-oriented terminology and use of extensible objects, greater interoperability can be achieved based on following these guidelines:

- Promote the ability to exchange requests between objects
- Allow the ability to enable objects to request services of other objects
- Ensure that this is regardless of the language in which the objects are defined, their physical location, their hardware platform, operating system, or DBMS

Distributed interoperable objects are objects that support a level of interoperability beyond the traditional object boundaries that are imposed by programming languages, data models, process address space, and network interface. A key objective in providing interoperability and distributed object management is a dynamic and open information universe that ensures scalability and effectiveness for remote data access and delivery.

Scalability within this context refers to the ability of distributed object management services to be scale rapidly when delivering information. The system implementation would be to deliver a set of objects from a variety of data sources to a large set of consumers. A large scale implementation needs to make sure that there is heterogeneity in the hardware platforms, storage locations, address spaces, operating systems, data models and programming languages, as the information sources continue to evolve.

Within this information environment, the architecture strategy is to allow abstractions of distributed interoperable objects to be captured and utilized by the distributed object management services to schedule and control the remote data access and delivery. The role of remote data access and delivery is to be able to use distributed object management services to incorporate the constant information evolution, the large increases in network traffic, to conduct distributed query optimization and execution strategies.

It is important to note that within the upcoming net-centric environment, data access will be across widely distributed and highly autonomous information sources. This imposes significant new challenges for distributed object management. This means that:

- There will be semantic and performance issues that arise due to the heterogeneous nature of the data sources
- There is a rapidly growing information base, since the amount and diversity of online information available across networks is exploding

With regard to database transactions and user expectations, information system owners have tried to solve their issues based on query optimization and evaluation strategies. These have mostly been in centralized, parallel, and tightly coupled distributed environments. However, as the amount of unrelated data grows, users today are faced with increasing difficulty in collecting, processing, and integrating information effectively and in a timely fashion.

Information system owners are realizing that as the scale and rate of change for online information continues, the architecture model that allows user-initiated, comprehensive searching is no longer sufficient. Within distributed information systems, the way to improve query responsiveness would be to build the system for tomorrow's needs. This would allow the following processes:

- Remote data access and delivery must be combined with user initiated comprehensive searching
- Address the interoperability demands of the information delivery system with source-initiated dissemination of relevant information
- Migrate from pre-established facility locations and communication endpoints to a broader architecture that allows anytime access, anywhere on the globe
- Ensure that the wide-area networks of are able to provide access to information sources, intermediate sites and communication links in a rapid manner.

## REVIEW OF CHAPTER GOALS

The goals of this chapter were:

1. To provide a broad and objective definition of Interoperability:
   - Address how does interoperability definitions differ from organization to organization.
2. To address what it means to have different interoperability types:
   - Describe what technical, operational, data, syntactic, and semantic interoperability are all about.
3. To introduce the definition of coupling and define loosely-coupled systems:
   - State why it is better to have systems that are loosely coupled, since close integration of systems cause a number of challenges when systems become more large-scale and distributed.
4. To discuss how interoperability can be measured in an objective manner within the enterprise:
   - Discuss the LISI profile, its PAID attributes and its scale from 0 to 4.
5. To provide an understanding of different architecture strategies that lead to greater interoperability:
   - Discuss how use of interfaces and layers and use of technical standards are important for large organizations.
6. To discuss challenges of ensuring interoperability among large scale distributed systems:
   - As systems become more diverse and distributed in nature, we have to find means for ensuring interoperability within a large enterprise.

## REFERENCES

Department of Defense C4ISR Interoperability Working Group. (1998) *Levels of Information Systems Interoperability (LISI)*. Washington, DC: Author. Retrieved from http://www.defenselink.mil/nii/org/cio/i3/lisirpt.pdf

Department of Defense Chairman of the Joint Chief of Staff (CJCS) Instruction 3710.01B (2001). *Requirements Generation System*. http://www.dtic.mil/doctrine/jel/cjcsd/cjcsi/3170_01b.pdf

Department of Defense Joint Publication 1-02 (2001). *DoD Dictionary of Military and Associated Terms*. Washington, DC. http://www.dtic.mil/doctrine/jel/new_pubs/jp1_02.pdf

Faughn, A. W., (2002, October). *Interoperability: Is it Achievable?* Program on Information Research Policy. Cambridge, MA: Center for Information Policy Research, Harvard University.

Kasunic, M., Anderson, W., (2004, April). *Measuring Systems Interoperability: Challenges and Opportunities*. Technical Note CMU/SEI-2004-TN-003.

Morris, E., Levine, L., Meyers, C., Place, P., & Plakosh, D., (2004, April). *System of Systems Interoperability (SOSI): Final Report*. Technical Report, CMU/SEI-2004-TR-004.

North Atlantic Treaty Organization (NATO) Backgrounder. (2006, July). *Interoperability for Joint Operations*.

Office of the Assistant Secretary of Defense for Networks and Information Integration, Department of Defense Chief Information Officer. (2001). *Global Information Grid (GIG) Capstone Requirements Document (CRD), Flag Draft*. Retrieved from http://www.dfas.mil/technology/pal/regs/gigcrdflag-levelreview.pdf

Office of the Assistant Secretary of Defense for Networks and Information Integration, Department of Defense Chief Information Officer. (2003, October). *Joint Technical Architecture, Volume 1, Version 6.0.* Department of Defense Chairman of the Joint Chief of Staff (CJCS) Instruction CJCSI 6212.01C (2003, November). *Interoperability and Supportability of Information Technology and National Security Systems*.

Starr, S. H. (2004). *The Challenges Associated with Achieving Interoperability in Support of Net-Centric Operations*. Falls Church, VA: Barcroft Research Institute Proceedings.

# Chapter 4
# Net-Centric Operational Environment

## CHAPTER CONTENT

As you explore Chapter 4, it will cover the following topics:

- Basic Tenets of a Net-Centric Operational Environment
- Addressing Problems of Our Military
- Net-Centric Operational Context
- Exploiting Knowledge and Technical Connectivity
- Knowledge Management within the Operational Environment
- Case Study: Example of Future Net-Centric Environment

## CHAPTER FOCUS

In this chapter we delve deeper into the concepts of net-centric systems and provide characteristics of the upcoming net-centric operational environment. We define traits that are properly defines this type

DOI: 10.4018/978-1-60566-854-3.ch004

of real-time operational environment and its benefits. The problem of today's military is described and a shift to a new operational environment is mentioned to surmount current challenges. The chapter then illustrates the net-centric operational context and denotes each of the elements. We describe the fundamentals of knowledge management, knowledge sharing and distributed decision making. We address issues such as collaboration within communities of interest, technical connectivity and network management principles. We define different knowledge domains and sharing of knowledge across the net-centric environment. We then provide a case study of an example, future net-centric environment that has been presented within a DoD joint forces document.

## BASIC TENETS OF A NET-CENTRIC OPERATIONAL ENVIRONMENT

Now that we have discussed the concepts of Net Centricity and Interoperability, it is important to discuss in more detail, what it is like to be within a Net-Centric Operational Environment (NCOE). This operational environment pertains specifically for today's U.S. military. However, this type of operational environment can be modeled for military planners for countries around the globe. This operational environment can also be used by civilian industries based on a technology transfer from today's military to commercial corporations.

With regard to harnessing the latest technology, NCOE allows us to explore a number of information aspects that fully enable upcoming technologies and innovations. This includes Communities of Interest (COIs), the ability to share and manage knowledge effectively, and an operational matrix that provides distributed decision making.

From a military and battlefield perspective, the following elements can be attributed to be part of an NCOE:

- Gain Information Superiority First
- Allow Deep Sensor Reach
- Gain Shared Battlespace Awareness
- Incorporate a Faster Decision Cycle
- Ensure that there is a Unity of Effort
- Conduct "Joint" Operations at the Tactical Level

Within the combat theater, NCOE then leads to the following characteristics for future warfare:

- Rapid Change
- Speed of Command
- Self-Synchronization of Actions
- Interdependent Operations

### Benefits of a New Net-Centric Operational Environment

Based on the DoD net-centric vision, the valued benefits from an NCOE include an information environment that addresses networking, sensing, information sharing, agility, risk tolerance and innovation. These principles allow the DoD to achieve a networked "joint" force that is linked and synchronized in time and purpose. The military aims to leverage technology to create new synergies in the way DoD trains, organizes and leads more effective and efficient military operations.

The NCOE concept from the military and Joint Forces combines areas of knowledge management, information assurance and network management principles. NCOE allows Joint Forces to possess a globally accessible platform of data and information.

Some resulting benefits from conducting net-centric warfare include the following:

- Networked forces can consist of smaller-size units that can travel lighter and faster, meaning fewer troops with fewer platforms. They can carry fewer supplies and can perform a mission at a lower cost more effectively.
- Networked forces can fight using new tactics. Because networking allows soldiers to keep track of each other when they are out of one another's sight, forces could move forward in Iraq spread out in smaller independent units, avoiding the need to maintain a tight formation.

A military organization derives a large number of benefits from operating within this new NCOE context:

- It requires fewer troops and less number of equipment
- The enemy has a harder time in effectively attacking a widely dispersed formation
- Combat units can cover much more ground, because they do not have to maintain a formation or slow down for lagging vehicles
- Knowing the location of all friendly units reduces fratricide during combat operations
- The ability to swarm the enemy allows attacks to be directed straight into the heart of an enemy command structure undermining support by operating from the inside, rather than battling only on the periphery
- The sensor-to-shooter time is reduced. Soldiers in the field have the capability to conduct an "on site analysis" of raw intelligence from sensor displays, rather than waiting for return analysis reports to arrive back from Continental United States

However, the central assumption for conducting future warfare in this manner is that the net-centric environment, allows disparate applications to communicate flawlessly with each other over the Internet. This is not an easy assumption to make. The DoD is fast changing their entire infrastructure to prepare the critical Global Information Grid so that these military strategies can become a reality. The infrastructure needs to be also extended to the enhancement of homeland security against all threats, foreign and domestic and the improvement of industrial enterprise efficiencies so that net-centric warfare can be achieved.

## ADDRESSING PROBLEMS OF OUR MILITARY

According to DoD Networks and Information Integration (NII) and Joint Forces Command, the problems of the military can be summarized and paraphrased to reflect the current sentiments of military experts.

- **Long-term Operating Challenges** - The Joint Force in 10 to 20 years will operate in an environment that is increasingly complicated, uncertain, and dynamic. The military will need to employ a number of asymmetric strategies to thwart potential adversaries. The proliferation of advanced weapons and information technologies throughout the globe will create additional stresses on all elements of the force.
- **Interoperability Challenges** - Future operations in defense of the nation, require an increased level of joint integration, as well as integration with other federal agencies, state organizations, and coalition partners. The current state of human and technical connectivity and interoperability within the joint forces is inadequate to achieve the levels of operational effectiveness and efficiency necessary for success in the new net-centric operational environment.
- **Information Needs for Joint Forces and Coalition Partners** – The joint forces and coalition partners need to have trusted, accurate and timely information so that appropriate knowledge can be shared to make superior decisions in an assured environment, sorting through large quantities of operational data
- **Approach to Warfare and Conflict Resolution** – Military experts assert that fundamental changes are necessary in how we approach warfare and our approach to conflict resolution. People will remain the centerpiece of the new environment, but changes within today's society, the overall global security environment, and rapid advance and proliferation of technologies will require an ongoing transformation of the military. The new approach states that the military strives to be more effective by using whatever resources are available at the moment. This would include a combination of intellectual prowess, available technology, facility and environmental constraints.
- **Net-Centric Challenges** – The DoD realizes that even if a grand vision of the future is created, and policies and doctrines are written, the difficulty in achieving a net-centric environment lies in two critical areas: Knowledge sharing and knowledge management amongst people and personnel with the commands, and technical connectivity and interoperability within the military environment.

## Current Platform-Centric Environment

We should recognize that the current posture for our joint military forces is a platform-centric environment at the echelons below the military headquarters level. A platform-centric environment means that individual and autonomous systems are brought together in a rigidly structured fashion to accomplish a mission.

A platform-centric environment creates barriers to the flow of information across joint forces and its mission partners. Each partner uses their own system-specific platform that generates data using components that are supported by dedicated command or organizational support elements. Each platform has optimized their processes to support only their particular systems. The systems in a platform-centric environment lack horizontal integration with other systems, creating stovepipes of data and information. For the joint forces, platform-centric integration is done in a centralized command center supporting higher echelons.

In a platform-centric environment, the quality of information along with the content, speed and format is dictated by formal requirements generation and fulfillment processes. These requirements employ centralized and functionally specialized information management, information technology processing, and acquisition practices.

*Figure 1. Net-Centric Operational Context*

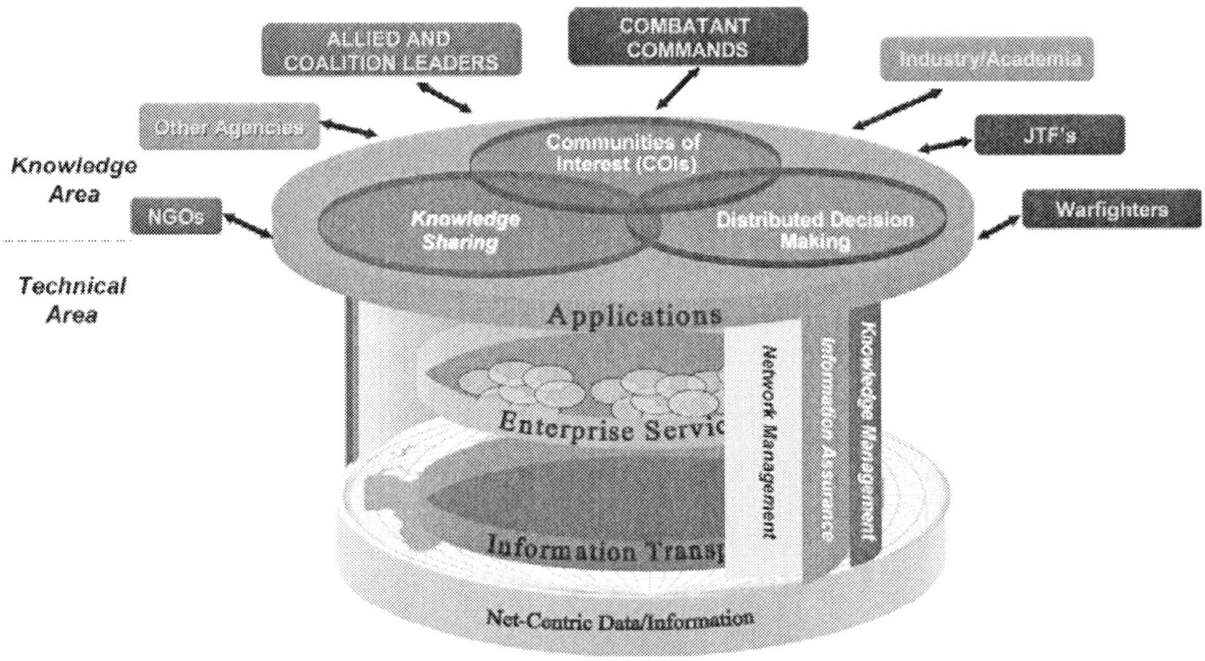


Since platform-centric environments tend to not be able to work well with other comparable systems, this causes delays and communication headaches for the soldier. This may then reduce the effectiveness and efficiency of the military forces. This type of approach is inadequate and will not work for the future. The future state of the military cannot be saddled with social and technical barriers to the flow of information. This then prevents tactical level integration of capabilities resulting in reduced effectiveness and efficiency.

## Fundamental Shifts in Operational Environment

In answering the military problems and challenges, DoD planners have realized that a fundamental shift in the operational environment is necessary. This shift has to be in response to the changing character and conduct of warfare and conflict resolutions. Military forces are increasingly being put into unfamiliar situations where the situation is complex, uncertain, and the operating environment is rapidly changing. In this type of environment, soldiers need to be able to rapidly integrate varied, dynamic, and often unanticipated sets of capabilities. These capabilities can be drawn from, across and beyond the joint military forces and coalition partners, to achieve the mission objectives.

This fundamental shift can be depicted in the new NCOE, where the net-centric environment reduces the existing structural impediments and enables the military commander to successfully meet his mission objectives. This new operational environment promotes the flow of information and reduces the friction of adjusting military joint forces and partner capabilities to new tasks and missions.

## NET-CENTRIC OPERATIONAL CONTEXT

Now that we have discussed the new NCOE thought process, we want to highlight the type of integrated environment that this signifies.

Figure 1 provides an illustration of an operational context that depicts each of the elements of an NCOE. The operational context integrates the functions of knowledge management, network management and information assurance within the NCOE. The primary intent of the NCOE is to provide the warfighter or military support personnel with a seamless capability to collect, create and use actionable knowledge in an operational context.

Each of the components such as knowledge management, knowledge sharing, distributed decision making, enterprise services, application and information transport layers needs to be integrated thoroughly within the context of the DoD Net-Centric Data Strategy.

If you are within an NCOE, you need to have support throughout the spectrum of national, strategic, operational, and tactical environments. The environment's capabilities need to support a set of users that can operate in austere, hostile environments, with minimal bandwidth and high latency. NCOE capabilities also should be able to operate under restrictive conditions in all phases of deployment such as mobilization, movement, employment, sustainment, and redeployment. The idea is that deployed forces have to allow for intermittent connectivity and their reconnection requires synchronization or re-synchronization with the core infrastructure.

## EXPLOITING KNOWLEDGE AND TECHNICAL CONNECTIVITY

The net-centric operational environment aims to exploit key parameters such as -- shared knowledge, collaboration, technical connectivity, distributed decision making, and network management principles. These parameters can thereby used to foster increased mission effectiveness and efficiency.

### Shared Knowledge and Collaboration

The latest information revolution depicted within today's information enterprise has created new ways in communicating within the military and their partner organizations. The ability to share information and collaborate allows a joint understanding of the operational environment.

Within this military context, collaboration is defined as a joint problem-solving exercise that achieves a shared understanding, is able to make a decision, or create a product across the joint forces and coalition partners.

Shared knowledge and collaboration is used to:

- Ease the manner with which information is accessed and updated
- Promote awareness of a common knowledge base
- Build a common understanding among the military forces and partners
- Support decision-making and synchronize activities

Participation in a collaborative process is an important right and responsibility that requires training, experience, and the confidence to interact effectively. It is important that military forces are prepared to exploit the information and actionable knowledge that is made available. They also have the ability to update this knowledge base by adding information gained from previous experiences, as well as adding real-time, tactical information.

## Technical Connectivity and Infrastructure

As part of the net-centric vision in moving toward a target state, the DoD and military is focusing on the realization of a robust end-to-end network infrastructure that is characterized by the Global Information Grid (GIG) initiative. The success of the GIG and related DoD initiatives is vital in building the technical architecture and foundation of the NCOE.

Within this environment, users need to be connected with adequate resources to allow reliable, real-time, continuous access to enterprise information and services. The NCOE does not imply that there are infinite resources, but the environment allows all organizational units to manage available resources to meet changing mission needs.

Traditional technical and network investments have centered on Command and Control requirements for deployed nodes. Transitioning from a platform-centric environment, which has been part of the command-and-control hierarchy, requires a sea change in internal and external organizations and policies. The new NCOE has to allow sharing, promote awareness and understanding, and foster superior decision-making based on a synergistic application of force capabilities. Agile operation of military forces requires enabling both formal and informal collaboration, and the ability to establish and utilize relationships with every mission partner.

At its heart, NCOE is a social construct that is supported by an advanced information infrastructure. The total capability of an NCOE is greater than the sum of the knowledge areas and technical connectivity. To understand the relationships between knowledge and technical connectivity, it is important to not regard information as integral to the physical or technical infrastructure, and not be tightly coupled to applications. Within the NCOE, information is posted to shared spaces that can be accessed by both anticipated and unanticipated users. This can happen through the use of loosely-coupled applications that employ smart, pull-based architectures.

## The Focus of Distributed Decision Making

In this new operational environment, it is expected that the decision-making process will be radically different from the past. Decision-making in a knowledge sharing environment can be heavily influenced by a dynamic, collaborative process. However, for military decision making, collaboration lacks individual accountability. Collaboration can inform a decision-maker but should not remove the decision-making responsibility from our military leaders.

A distributed decision-making process facilitates better planning and execution by enabling all partners to share common mission objectives. This helps in synchronizing the operation and organizes tasks for optimal efficiency. Distributed decision making allows military forces to:

- Better interpret situations and problems
- Identify candidate actions
- Formulate evaluation criteria
- Decide what to do
- And execute the final decision

It is expected that within the NCOE, every authorized user does not need to be an expert in the mission that he faces. Yet they can collaborate and make decisions based on knowledge provided by a subject-matter expert or several subject matter experts. The idea is that the NCOE will provide every authorized user with assured, high-bandwidth access:

- to subject matter experts via video-conferencing, e-mail, and other means
- to high quality analyses and recommendations for the mission area
- to other relevant information such as discussion databases and chat rooms of colleagues

### Network Management Principles

Network management principles within this operational environment address people, technology, processes, policies and capabilities. These principles help to effectively operate the systems and networks, their configuration, availability, performance, manageability, and enterprise connectivity. It is expected that network managers oversee the operation of the information transport, services, applications, computing infrastructure, spectrum, content staging, and other capabilities as necessary to ensure the proper functioning of the network at their respective levels of responsibility.

For conducting network management, tools will allow managers to synchronize their actions. Smart tools can be created in the form of applications and services that is used to speed and automate network management functions. With regard to distributed data, a clearly written set of metadata descriptions and directories will be used to identify services that can be retrieved by authorized users.

It is expected that network management tools can interface with data links of legacy systems, with packet-switched Local Area Networks (LANs) and high-capacity Wide Area Networks (WANs) through network gateways. Upcoming mobile technology also adds radio frequency and wireless systems that can be connect with satellite-based data networks. At optical gateways, radio frequency data-links or voice can converge with other data signals such as terrestrial voice, data, and imagery traffic.

Network Management requires further study to figure out how operational traffic is to be prioritized and given precedence based on a set of dynamic criteria. A software-driven policy rule sets will be used to determine which traffic of what type gets access to which ports under what conditions. Network Management also needs to mitigate the effects of network degradation, outages, and attacks. When such problems arise, a distributed, global network management community needs to act quickly and collaboratively to determine how the information flow can be optimized and who must take action accordingly.

## KNOWLEDGE MANAGEMENT WITHIN THE OPERATIONAL ENVIRONMENT

Currently, our military forces lack the capability to provide a dynamic, knowledge management environment that can be used by mission partners, such as other government agencies, multinational, and coalition agencies, non-governmental organizations (NGOs), industry, and academia. Within the NCOE context, the vision is to provide the following:

- Ensure that the right information is:
    - Available to the right person

- At the right time
- In the right context

Although a number of definitions exist for knowledge management throughout the public and private sectors, they generally refer to similar processes. Knowledge management is a systematic process that allows you:

- To discover, select, organize, distill, share, develop and use information
- All in a cohesive knowledge base or domain

Knowledge management stems from the premise that an organization's competitive advantage is based on its own collective knowledge. We all know that a substantial portion of this knowledge is not written down because it stays within individual workers within the organization. Harnessing the minds of people is the key to organizational knowledge management. This includes the collective memories, experiences, and the inner knowledge of what works and what doesn't. Whenever an experienced person retires from an organization, his departure may reduce that organization's collective knowledge. This then requires active management oversight to build the collective knowledge back up to meet the needs of the next set of challenges.

In an NCOE environment, the future state requires a paradigm shift from a "need to know" to a "need to share" orientation. The idea is to support dynamic organizational constructs that promote decentralized decision-making in a fluid environment. This is required not only to connect to existing command-and-control elements, but also to sensor systems and weapons systems.

Military forces require the capability to leverage uncertainty. The idea is to prepare for dynamic adaptability that can be achieved via knowledge sharing and net-centric operations. Future conflicts will likely be contests in adaptability where the winning force will be that force which adapts itself more effectively to the situation at hand. This will lead to faster conclusions for completing the battle or winning the peace.

## Description of Knowledge Domains

We want to introduce the concept of knowledge domains. There are four domains that describe the state of information and knowledge and how one can be derived from the other.

- **Information Domain** - is defined based on where information exists. This domain has a dual nature that consists first, the information itself and second, the medium by which we collect, process, and disseminate this information.
  - Characteristics of the information domain include:
    - **Information quality** – this includes completeness, accuracy, timeliness, relevance, and consistency
    - **Information Distribution** – this includes range, sharing, and continuity
    - **Information Interaction** – this includes the exchange or flow of information
- **Physical Domain** – is the domain where military forces would be moved through time and space. For the military, this spans land, sea, air, and space environments. This is where physical military platforms exist, and where the network and communications infrastructure reside that connect the platforms.

- **Social Domain** - is shaped by the knowledge management specifics of language and symbolic communication among human beings. This is the domain where individuals will interact with each other and is influenced by the collective knowledge within the group's mind-sharing. Face to face communication and knowledge transfer is by far the best way to share knowledge. Most face-to-face communication is actually non-verbal that is relayed through facial expressions and other body language, This means of communication enhances the verbal and written methods for knowledge dispersal.
- **Cognitive Domain** – is the least tangible of the four domains since it exists in the minds of human beings. The cognitive domain however, is influenced by a number of factors such as training, experience, public opinion, and situational awareness. The cognitive domain is where decisions are made and is directly related to intellectual capabilities and developmental levels.
    - The characteristics of the Cognitive Domain include those that affect individual and organizational decision-making
    - This includes attitudes, opinions, beliefs, values, and understanding

## Knowledge Sharing Mechanisms

We would like to mention that knowledge sharing mechanisms are vital to the military. Actionable knowledge, one which allows military leaders to make decisions cannot be ascertained if the knowledge sharing format or medium is inadequate. The outcome of knowledge sharing activities should be to work toward a common knowledge base, which is then used for the purposes of decision making and tracking ensuing actions.

Actionable knowledge is typically a combination of both implicit and explicit knowledge. Implicit knowledge is what individuals have within their mind, which is innate and allows them to come to independent conclusions. Explicit knowledge is what is found in books and information databases. A combination of these elements then produces intelligence that can be catered to military commanders as actionable knowledge.

Knowledge sharing is more than information sharing since it is based on how individuals react to a set of data or information. Although individuals interpret and process knowledge differently, they share knowledge using a finite number of techniques, such as through agreed formats and by training to specific methods and standards. Knowledge sharing also includes glossary information and data ontologies to fully explain the information jargon that is accessed and received. The ability to combine metadata explanations to the data that is transferred and provide contextual meaning allows knowledge sharing to be a far more powerful activity than merely sharing information.

Knowledge sharing within the military for official purposes is usually referred to as collaboration. Military forces have to collaborate to perform joint problem-solving for the purpose of achieving a shared understanding. This collaboration can be either formal or informal. However, collaboration efforts may result in actionable knowledge that is then acted upon by individual decision makers.

There is a variety of knowledge sharing technologies in today's enterprise. This includes the ability to use videoconferencing, shared whiteboards, audio teleconferencing, chat rooms, discussion databases, and email. Knowledge sharing can be real-time, which means that it occurs simultaneously, or in a synchronous manner. It also can be at the behest of the user, which means that it occurs asynchronously.

For synchronous, real-time communications, video-conferencing is usually a good choice. However, for asynchronous collaboration, email is a popular tool of choice. We know that no collaborative tool is ideal for every situation. In today's enterprise, personnel use a combination of synchronous and asynchronous tools to knowledge share effectively. Within the new NCOE context, a host of new collaboration technologies are being devised to address the challenges of net-centric warfare.

## CASE STUDY: EXAMPLE OF FUTURE NET-CENTRIC ENVIRONMENT

This case study has been excerpted from the DoD document, "Net-Centric Environment Joint Functional Concept", Version 1.0, April 7, 2005. This text is provide as a reference at the end of the chapter. The information that follows has been reproduced verbatim from the DoD document so that the example net-centric scenario is kept intact.

As it states within the document:

*This example is illustrative only and is intended to provide the reader with an understanding of how the U.S. military might function in a future Net-Centric Environment (2015-2025). It is to be used only within the context of this functional concept.*

### Background Scenario

In August 1999, strong earthquake tremors struck Turkey and caused significant damage.

The North Anatolian Fault that caused these tremors stretches to Istanbul beneath the Sea of Marmara. With the help of the U.S., NATO and the European Union, Turkish officials developed a robust, survivable network called Network Respond. Network Respond consists of numerous connected networks, strategically placed sensors, and databases to provide area data and information. The network uses a number of redundant communication and power systems and dispersed archives to protect against the effects of another catastrophic earthquake. Completed in 2020, this network connects the major cities that lie on this fault line through key nodes, which are interfaced with people and sensors in cities' high rise structures, hospitals, fire fighting stations, electrical, and telephone systems, transportation system, water and sewer systems, and oil refineries.

In 2022, U.S. Joint Forces are operating in a mature Net-Centric Environment. Knowledge and technological advancements have resulted in an unprecedented ability of joint forces to share awareness and create shared understanding. U.S. Joint Forces are able to operate seamlessly at the tactical level in dynamic Communities of Interest (COI) which can access the numerous resources including Network Respond. This agile force can rapidly combine capabilities from different services at the appropriate level, to efficiently accomplish an increased range of missions. This is the ability to achieve constructive interdependence, and it is the norm—not the exception.

### The Networked Setting

During the period 2010 to 2025, U.S. Joint Forces relationships with U.S. civilian law enforcement agencies, the Department of Homeland Security and appropriate agencies within the intelligence community have grown significantly. U.S. Joint Forces have also maintained very strong military relations with NATO and other foreign militaries. Multinational Standard Operating Procedures (SOPs) and

Tactics, Techniques and Procedures (TTPs) have been developed and are in use daily. Multinational training events have become commonplace, and foreign militaries have joined with the U.S. military in developing common interfaces, policies, and protocols. Individuals are able to filter, structure, and visualize shared data and information in meaningful ways. Initiatives to enable multinational information sharing are providing the capability for U.S. and Allied militaries to share data and information transparently and effortlessly.

In addition to improved multinational interoperability, many countries have paid particular attention to the need to develop seamless access to critical humanitarian information. The United Nations (UN) established a network to coordinate Humanitarian Assistance/Disaster Relief (HA/DR) among member nations and external groups such as participating International Organizations (IOs) and Non-Governmental Organizations (NGOs). This network, called the International Humanitarian Relief Network (IHRN), incorporates common interfaces, common standards, and common protocols (including security protocols) to allow all recognized participants the ability to access required information to support the range of required functions (e.g., medical, logistics, protection, engineering, etc.) through their organic networks. Numerous exercises have been held over the years using IHRN, and as a result, SOPs and TTPs have been developed for use by all participating countries and organizations. Participants have developed the required network interfaces, and have become accustomed to trusting one another through frequent posting and sharing information.

## Situation

At 4:15 a.m. on 25 March 2022, the Anatolia fault line ruptures causing a massive earthquake registering 8.2 on the Richter scale. The city of Istanbul is near the epicenter of the earthquake and suffers massive damage and destruction. The cities of Izmit, Golcut, and Bursa are also on the path of the fault and suffer significant damage and casualties. Aftershocks also contribute significant damage to the area. Combined, these cities have over 150,000 dead, 400,000 injured and 600,000 people homeless.

Due to the magnitude and severity of the earthquake damage, the Turkish government officially requests support from the UN and NATO. The UN responds by directing its Office for the Coordination of Humanitarian Affairs, in Geneva, to facilitate UN242 sponsored humanitarian support. NATO stands up a Combined Joint Task Force (CJTF), led by U.S. European Command (USEUCOM), and begins synchronizing its activities under the auspices of the Turkish civilian emergency management agencies and the Turkish General Staff. In response to the earthquake disaster, the CJTF launches Operation Combined Response to provide humanitarian relief and coordinate relief efforts supporting the areas in Turkey devastated by the earthquake.

Numerous IOs and NGOs respond to the Turkish appeal for help. Among these organizations are the International Federation of Red Cross and Red Crescent Societies (IFRC), CARE, and World Relief. The Organization for International Relief and Support (OIRS), a Syrian-based group chartered in 2015, also participates in the earthquake relief effort.

The U.S. Federal Government is inundated with offers from States and U.S. agencies to support Operation Combined Response. Many States have stand-by quick reaction Emergency Response Teams (ERTs), Urban Search and Rescue (USR) teams, and equipment that immediately deploy to Turkey.

## Execution

The headquarters of the CJTF is formed from a standing EUCOM element supported by a pre-established collaborative network consisting of both standing and dynamic communities of interest. Permanently assigned CJTF personnel are cross-functionally organized and have established strong, standing relationships with other functional experts within the military and humanitarian relief communities. Because of this, the CJTF is able to stand up very quickly, and while deploying to a location near Eskisehir, Turkey conducts seamless en route planning, coordinating, and directing of tasks and activities for Operation Combined Response. The CJTF consists of the U.S., Bulgaria, Greece, Italy, U.K., Canada, and France. Non-NATO members such as Israel, Japan, Russia, Austria, and Switzerland also immediately begin coordination with the CJTF and deploy ERTs and USRs to provide assistance as necessary.

The CJTF commander immediately establishes an interactive and distributed collaboration session with all of his commanders, their primary staffs, the State Department, US Embassy, the Defense attaché, and key IOs and NGO participants who enter the IHRN network to begin mission analysis and COA development. All CJTF participants are granted access to the Operation Combined Response COI to allow the sharing of information they will need to conduct this HA/DR support operation.

The CJTF is able to immediately access Network Respond and display realistic visualizations of structural damage to key buildings and the operational status of the area hospitals, firefighting stations, and police stations from protected archives of existing databases constructed, populated and initially updated by the Turkish civil authorities.

Seventy percent of the Network Respond sensors placed in strategic locations survived the earthquake and are able to send data regarding the location of casualties. Network Respond information quality and availability is assured through the use of automated network management tools designed to maximize the accuracy and reliability, utility, and integrity of data and information.

Turkey provides a collaborative team to the CJTF that functions as an information "broker" and uses various software tools to tag Turkish source data and information for specific content and releasability to respective nations and organizations participating in Operation Combined Response. This is done based on pre-determined COI data standards, supporting a framework with multiple levels of security.

Through a standing IHRN COI, all participating IOs and NGOs that had previously supported UN-led operations through the IHRN, are able to access the network and get the same data and information (situational awareness) that is available to the CJTF.

Those IOs and NGOs that did not participate in developing IHRN are able to rapidly connect to the IHRN and gain access as full participants in the COI. Intelligent user defined agents assign each of these organizations a level of participation in the COI commensurate with their roles, authorities, requirements and risk profile. By operating in a Net-Centric Environment, ERTs and USR teams are able to collaborate with CJTF units, other response teams, and all pertinent relief organizations, synchronize their actions, quickly deploy to areas where people are potentially trapped inside buildings and execute immediate search and rescue actions. All organizations responsible for casualty activities automatically post casualty updates, allowing network participants to access near real-time information on current casualty locations, status, severity of injuries, and availability and location of nearest ERT and USR teams and equipment, supplies, current on-site conditions, and status of casualty logistical/medical support infrastructure.

On March 27, two days after the earthquake, a massive car bomb explodes outside the Hotel Bandora in Ankara, approximately 250 miles from the Istanbul area relief effort. The bomb kills 10 key members of the Greek Cypriot-controlled government and 20 high ranking members of the Turkish contingent who are attending a Cyprus Unification Seminar. Forty-five bystanders are also killed and 150 individuals are injured in the explosion. Shortly after the bomb explodes, the terror group Al Shalib Hurstat claim credit for the incident citing their disapproval of the Cyprus Unification Seminar and threatening more terror activity if the unification efforts continue.

The CJTF is given the additional mission of providing force protection, and support to help the Turks locate and neutralize the terrorist cell responsible for the bombing. This new mission is designated Operation Stomp Out. Taking advantage of the shared situational awareness and understanding achieved during Operation Combined Response, the CJTF immediately establishes an interactive collaboration session with all commanders and primary staff members to update the situation and begin mission analysis.

The CJTF establishes the Stomp Out COI to assemble all relevant information related to active and inactive terrorist cells operating in and around Turkey. The CJTF Commander tasks this COI to develop a recommendation on the likely terrorist cell responsible for the bombing, its disposition and likely location. To accomplish this task, the COI immediately realizes that it needs the means to assemble and analyze all data and information related to terrorist cells, terrorist supporters suspected of planning and/or conducting terror in the Area of Responsibility (AOR), local leaders, previous terrorist incidents, and responsible parties. Therefore, the COI quickly expands to include not only the organic CJTF ISR assets but also the Turkish Liaison Officer and his resources, the EUCOM J2, CENTCOM JTF-CT, the Defense attaches at the American Embassy, and a North Atlantic Council Counter Terrorism Force that was established in 2008. The network allows the CJTF to quickly and easily reach back to other assets without increasing the footprint of the forces required to support operations in Turkey. This reduces the time and resources needed to bring additional information sources and counter-terrorism capabilities to bear on the problem at hand. Because of the nature and location of the event, the Turkish liaison officer is identified as the COI leader.

There is a great deal of data and information pertaining to Ankara and its surrounding areas on Network Respond and the Turkish government allows the CJTF access. CJTF mission partners' access is based primarily on operational roles, as delineated by the CJTF and as stipulated by the COI leader.

A logistics COI is established that plans for acquiring and managing the resources needed to provide logistical and medical support to Operation Stomp Out. This dynamic COI provides peer-to-peer connectivity for logisticians in each unit supporting the operation, EUCOM logistics planners, and U.S military component logistical planners. The logistics COI conducts collaboration necessary to support the new operation allowing this COI to assess the logistical status of Operation Combined Response, identify the support requirements necessary to respond to the event in Ankara, and analyze the in-transit status of supplies. This provides the means to develop a comprehensive recommendation to the CJTF to redirect certain critical support from Operation Combined Response to Operation Stomp Out.

The NATO Rapid Reaction Force (RRF) is placed under the operational control (OPCON) of the CJTF. In 2022, the RRF consists of a Brigade Combat Team (BCT) with battalion-sized combat units, military intelligence, engineer units, military police units, and signal/communication units as well as RRF level support units. The RRF planning element is able to tie into the COIs for both Operation Combined Response and Operation Stomp Out.

The RRF tasking in Operation Stomp Out allows its units appropriate role-based access to network operational data and information. The plan cell automatically subscribes to any data or information posted on the network related to terror activities, terrorist supporters, weapons, and then further processes this information on its tactical network.

Smart agents alert RRF units with mission specific information as determined by individual users. Individuals further selectively filter this information based on their specific information needs. On March 28, a Turkish doctor working in an OIRS medical facility in Izmit, reports overhearing a conversation of one of her coworkers that leads her to believe that the coworker and possibly other OIRS members have ties with Al Shalib Hurstat.

This information is reported to the Turkish government, which directs the information be immediately sanitized, tagged with appropriate security labels and posted The report is fused with other data and information related to Al Shalib Hurstat and OIRS and, as a result, the OIRS's access to information on the network is quickly restricted due to a perceived security risk. However, OIRS retains access to local non-sensitive humanitarian relief data and information.

Concurrently, numerous other data and information related to terrorists are posted by various mission partners in Operation Combined Response and Operation Stomp Out, intelligence agencies, and sensors. Local inhabitants who are on the ground providing assistance and relief also provide key information to members of CJTF. These Human Intelligence (HUMINT) reports are automatically tagged and posted as they are reported.

The Stomp Out COI has subscribed to information related to suspected terrorists in the AOR. As a result, the COI automatically receives the OIRS report and begins the collaboration necessary within the intelligence community to fully analyze the data regarding all information relevant to this situation. The COI collaboration is focused on assessing the fused data/information that is coming in to provide an update to CJTF and the RRF's situational awareness. Based on the comprehensive collaboration amongst the COI participants and the new information related to Al Shalib Hurstat, the COI ascertains that the terrorist group Al Shalib Hurstat is indeed responsible for the bombing and that these same terrorists are assembling in the city of Kayseri about 250 miles from Syria.

The RRF immediately deploys the BCT to Kayseri; however, the BCT has little information on the city's design, layout and transportation network. Though available, satellite imagery will not provide the details needed to fully plan a combat mission in Kayseri. The RRF commander considers a request to EUCOM to provide additional forces capable of providing detailed imagery of Kayseri.

One of the military units supporting Operation Combined Response is a U.S. Army Unmanned Aerial Vehicle (UAV) unit that is providing aerial support to locate and rescue casualties. The UAV unit has a platoon that can provide long range urban/MOUT aerial reconnaissance support and this platoon is not currently supporting Operation Combined Response. The UAV commander is connected to the network and has visibility of the situation unfolding. The UAV commander contacts the BCT commander and, after collaborating on the situation, offers his platoon as a quick solution to providing aerial reconnaissance over Kayseri. The mission change requires extra security for the UAV downlink sites, which the BCT is able to easily accommodate. Logistics clerks from both units use the CJTF logistics COI to arrange for delivery of supplies needed to support the new arrangement. Members of other functional areas also make appropriate adjustments to ensure that this important task is adequately supported.

The RRF commander has configured his information visualization system to track this type of development and informs the CJTF, EUCOM and the Turkish General Staff of the situation. Within hours the BCT receives meta-data tagged imagery with embedded geospatial data from the UAV platoon. The BCT in collaboration with units and COIs throughout the CJTF (including the Turkish General Staff and its civilian leadership) quickly exploits the information and develops a plan to strike the terrorists. The constructive interdependence achieved by the rapid tactical level integration of UAV,

BCT, and supporting COI capabilities allows the CJTF to successfully execute a mission that results in the capture of the terrorists.

## REVIEW OF CHAPTER GOALS

The goals of this chapter were:

1. To allow you to understand what a net-centric operational environment is all about:
    - Describe some of the benefits that make it different from today's operations.
2. To engage in a discussion on what types of problems our military is currently facing:
    - Address how the rigidity of a platform-centric environment can be made more agile by becoming net-centric.
3. To provide an illustration of the Net-Centric Operational Context:
    - Describe each of the elements configured within this context.
4. To promote the core aspects of sharing knowledge and assuring connectivity:
    - Address shared knowledge, collaboration, technical connectivity, distributed decision making, and network management principles.
5. To describe knowledge management principles within the operational environment:
    - Describe the information, physical, social, cognitive domains and knowledge sharing mechanisms.
6. To provide a case study example of a future net-centric operational environment:
    - Describe how a massive earthquake in Turkey is operationally supported and executed.

## REFERENCES

Alberts, D. S., & Hayes, R. E. (2005, April). *Power to the Edge*. Command and Control Research Program Publication Series. Available from www.dodccrp.org

Department of Defense, (2004, December). *Net-Centric Environment Joint Functional Concept, Version 0.95*. Washington, DC: Author.

Department of Defense Joint Staff Washington DC, (2005, October). *Net-Centric Operational Environment - Joint Integrating Concept, Version 1.0*.

Eleazer, B., Reeder, S., (2007, April). *Testing Concept of Operations (CONOPS) in DoD's Net Centric Environment Testing*. 8thAnnual NDIA Science & Engineering / DoD Tech Expo.

Gaetjen, Tom, Lt. Col., (2005, May). Army G6, *Net-Ready Key Performance Parameter*. DoD Briefing Presentation.

Lt. Gen. Charles E. Croom, Jr., U.S. Air Force. (2006, July). *Service-Oriented Architectures in Net-Centric Operations*. Retrieved from http://www.stsc.hill.af.mil

Moser, Scott, Lt. Col. (2005, November). *Interoperability Certification Process and NR-KPP Brief.* Washington, DC: Department of Defense.

Office of the Assistant Secretary of Defense for Networks and Information Integration, Department of Defense Chief Information Officer Directive 8320.02-G. (2006, April). *Guidance for Implementing Net-Centric Data Sharing.*

Office of the Assistant Secretary of Defense for Networks and Information Integration, Department of Defense Chief Information Officer. (2001). *Global Information Grid (GIG) Capstone Requirements Document (CRD), Flag Draft.* Retrieved from http://www.dfas.mil/technology/pal/regs/gigcrdflag-levelreview.pdf

Office of the Assistant Secretary of Defense for Networks and Information Integration, Department of Defense Chief Information Officer (2008). *Department of Defense Information Management/ Information Technology Strategic Plan 2008-2009.*

Operational Test Agency Commander's Net-Ready Key Performance Parameter Study Group, (2005, October). *Net-Ready Key Performance Parameter (NR-KPP) Study Report, Version 2.2.* Alberts, D.S., Garstka, J.J., & Stein, F. P. (1999). *Network Centric Warfare: Developing and Leveraging Information Superiority,* (2nd Ed.). Washington, DC: CCRP Publication Series.

# Section 2
# Transformational Perspectives

*This next section focuses on the transformation aspects of the net-centric future and addresses the book's objectives to provide perspectives and strategies as to how this affects the information enterprise. There are three chapters within this section.*

- **Chapter 5:** *Target State for Defense Information Enterprise*
- **Chapter 6:** *Net-Centric Military to Civilian Transformation*
- **Chapter 7:** *Healthcare Transformation in a Net-Centric Environment*

# Chapter 5
# Target State for Defense Information Enterprise

## CHAPTER CONTENT

As you explore Chapter 5, it will cover the following topics:

- Viewpoint on Managing Tomorrow's Information
- Transforming to the DoD Target State
- DoD Business Transformation Activities
- Defining the Defense Information Enterprise
- Reference Model for DoD Information Enterprise
- Net-Centric Transformation Goals
- Defense Information Enterprise Priorities

DOI: 10.4018/978-1-60566-854-3.ch005

## CHAPTER FOCUS

As the first chapter on transformational perspectives, this chapter provides you how the DoD and the military are moving toward a target state for the enterprise. The defense information enterprise has a particular set of goals and mission needs that stem from the overall defense transformation. A guest author perspective provides a unique window to the activity. The vision and mission of the business transformation has been stated, along with providing a proper definition of the defense information enterprise. A pictorial depiction of the reference model for the enterprise is then presented, and each element is further described. We provide a transformation perspective on the DoD information enterprise. We then discuss enterprise as it relates to a set of net-centric goals. These net-centric goals branch into a set of priorities that include data and services deployment, secured availability, communications infrastructure readiness, communications readiness and NetOps agility.

As we discuss Net Centricity and the ongoing net-centric transformation, we plan to provide perspectives as to how this mindset is affecting our military and civilian enterprise. The first section of this book provided an in-depth understanding of the concept of Net Centricity and defined important concepts such as information interoperability.

In this chapter we discuss the Defense Information Enterprise since this provides a tangible product of the DoD transformation activities. The idea is to define the target state of the Defense Enterprise – what are its basic tenets, and how it leads to a greater level of Net Centricity. Based on our earlier discussion on today's information enterprise and the various information forms and the decision making capabilities, we now branch into how this is configured for our nation's defense. We also provide a guest author perspective on the transformation and the defense information enterprise.

## VIEWPOINT ON MANAGING TOMORROW'S INFORMATION

Let's start by recognizing the information revolution that has been happening in the past several decades. In developing a perspective on managing tomorrow's information, we have to recognize the characterizations of today's information enterprise that we discussed in the previous chapter. Here are a set of actions that we are deriving for the upcoming future that abide by net-centric principles:

- **Harness the Decision Making Pyramid** – Enable the stepwise evaluation from data, to information, to knowledge, to intelligence, so that it leads all of us to more informed decision making
- **Leverage Today's Information Data Store** – As information is stored in digital format, the written text, and created images are proliferating throughout the globe that are all being kept within computer directories and database repositories. As these information data stores grow by leaps and bounds and become more far-flung, it is more important for us to find new ways to leverage this information.
- **The End of Hierarchical Command & Control** – The past mechanisms for decision making has been primarily hierarchical. This means that information was provided to the top of an organization, so that decisions were made and filtered down to the bottom of the labor force. This led to a command and control structure that kept information at the top and prevented it from flowing downwards. The end of this paradigm means that information would be more free-flowing and information sharing strategies are necessary to properly manage the entire enterprise.

- **The Advent of Automated and Matrix-Based Decision Making** – As Net Centricity takes hold, the idea is that decision making may become more automated and mechanized. This means that information and decision logic is input into computer networks that automatically make decisions based on the proper circumstance. This also promotes a matrix-based decision making ability that allows groups and multiple individuals to independently make decisions and change the outcome of the final solution.
- **Loosely –Coupled vs. Closely Coupled Interactions** – As government organizations and the industry marketplace start using similar set of information processes across the globe, it is best to adopt a more loosely-coupled information stance, as opposed to close coupling. This means that organizations remain independent of each other and information sharing mechanisms abide by information assurance rules to prevent private and critical information to be shared by everyone. Disparate systems should not be closely coupled during a systems integration phase since this leads to more system failures and becomes more cost-prohibitive.

These central tenets now can be set as goals to achieve for tomorrow's information enterprise. The rest of this chapter provides an understanding as to how the DoD and the military is keeping pace with the rapid change that is taking place within the information enterprise.

## TRANSFORMING TO THE DoD TARGET STATE

To lead the defense and military of the United States through this next century, the National Defense Strategy of the United States of America (NDS 2005, p.10) states the following:

"Transformation is not only about technology. It is about:

- Changing the way we think about challenges and opportunities;
- Adapting the defense establishment to that new perspective, and
- Refocusing capabilities to meet future challenges, not those we are already most prepared to meet."

This means that transformation is more than changing information processes it requires a policy and culture change to enable this strategy to take effect. The DoD Chief Information Officer recognizes that information is a critical enabler for ensuring mission success. A comprehensive strategy has been developed that includes information management and information technology, to provide a common understanding of the shared vision, mission and governing principles of the target state.

**VISION**: **Deliver the Power of Information** – An agile enterprise empowered by access to and sharing of timely and trusted information

**MISSION**: **Enable Net-Centric Operations** – Lead the information age transformation to enhance the DoD's effectiveness and efficiency

To manage tomorrow's information and use tomorrow's upcoming technology tools, DoD not only requires a strategy, but also the people's will to transform the business of the government agency to best serve the needs of the citizens of the country.

## TRANSFORMATION PERSPECTIVE ON THE DoD INFORMATION ENTERPRISE

*This transformation perspective is from an industry veteran who comments on the current state of our defense information enterprise. Current efforts are ongoing to transform our defense to a more flexible, adaptive enterprise and provide more intelligent decision making capabilities to the remote warfighter in the field. By Dan Adams.*

Mr. Dan Adams is a military subject matter expert and works for the U.S. Army conducting acquisition enterprise requirements, plans and programs, statutory and regulatory financial management issues. Mr. Adams retired as Lieutenant Colonel from the U.S. Army.

War fighting is dependent on information. Its most basic questions are informational in nature:

- "Where is my adversary?"
- "What is my adversary's force composed of?"
- "What are the weaknesses I can exploit?"

The ability to control the decision cycle, and seize and maintain the initiative, is fueled by accurate and timely information. War fighters understand this at their core, and substantial investments in "information fusion" are being made in the tactical and operational space.

DoD has a massive business enterprise, which is also fueled by information. But as big as the defense business enterprise is, it exists solely to support the war fighting function of the U.S. defense department. Its success is not measured on a balance sheet, instead it is measured in conflict, near conflict and prevented conflict. But the business enterprise confronts the "Fog of War" as surely as a practitioner in the tactical and operational space.

By law, the DoD is a federated enterprise, and the Departments of the Army, Navy and Air Force have traditionally been very zealous in the protection of their functions under Title X of the U.S. Code. Each service, and sub community within the service has grown their capability to manage information with the advances of technology. Over the past sixty years this has created a veritable Tower of Babel, with limited and expensive system to system inter-connections. Of course this is not unique to the DoD, since this is also endemic in the private sector for our very large corporations.

The case for change is clear. Operating the DoD Business enterprise in the seventh year of a war has pointed out that it must be capable of responding to asymmetric threats. The business enterprise needs to be able to turn inside the opponent's decision cycle just like a tactical operator. Twenty year weapon system development cycles are very good in planning for the "next war", but they are unresponsive to current war fighter needs. The emergence of highly lethal improvised explosive devices required ingenuity and drove the "rapid" adoption and fielding of Mine Resistant Ambush Protected (MRAP) vehicles. Yet that decision to field is only "rapid" when compared to the current definition of normal.

But just as advances in technology compounded a cultural problem, it also offers the solutions. As the dollar cost to share information falls, old barriers to information sharing weaken, and the pressures of being at war compound the need for situational awareness. Operating the enterprise in a sustained conflict drives home the recognition of the existence of end-to-end processes, and that there is only one enterprise. This is a cultural epiphany, not a technical roll-out.

Technologies alone can only provide efficiencies; technologies can never make an ineffective business process effective. Not surprisingly the first step in addressing the interoperability needs in a federated enterprise is a cultural step. Here are a set of thoughts to dispel the common ailments.

- "Drive out fear" - Focus on the free flow of information within the component. Information free flow within the department is not a component concern, but the very tools and structures required to achieve that flow serve the larger enterprise. By habituating leaders to the exposure of information, the environment conditions the way they respond to it.
- Ensure that information is always provided in its context. All information has a sell by date, after which its consumption is hazardous. It also has a myriad of other attributes that are dependent on the system or process that generated it.
- Focus on the quality of information provided to unambiguous external forcing functions such as the requirement for a "clean audit"
- Finally, here is an opinion with regard to secure and compartmentalized information. The free flow of information means that if the information you require to make a decision is available, then you can have it. But nobody needs everything.

Change along this axis enables a shift from a response with the least bad decision, permitting seizure of the initiative, and producing the opportunity for a decision that is closer to optimal.

## DoD BUSINESS TRANSFORMATION ACTIVITIES

At this time, the U.S. Department of Defense is conducting a thorough business transformation. The new model needs to allow the defense business enterprise to adapt and flex as decision makers deal with growing competitive pressures, changing regulations, and strategic and tactical course shifts. In this model, transformation is about reacting instantly and effectively to what today's modular, agile, technologically-advanced joint force demands. This new reality has given increased urgency to transforming the Department's business operations. Transforming these operations is no longer just a matter of achieving cost, schedule, and performance requirements. It's about business agility.

For the past several decades, the Defense Department's business model has been configured to support a military dependent on large-scale weapons systems and prepared for sustained, predictable battlefield engagements in specific parts of the world. The contemporary defense business model has to be different, to keep pace with the changing nature of the DoD's warfighting challenges.

### Embracing Change

The DoD has made significant strides in breaking down the cultural and systems barriers that hinder business agility. There is an increased need for tighter alignment of end-to-end business functions, better management visibility into operations, and a noticeable bias toward execution excellence. The current climate is to make measurable business improvements every six months, which is tied to an enterprise transition plan that drives progress.

Changing the cultural mindset has meant redefining defense business in terms of the customers it serves, rather than the functions it performs. Breaking down systems barriers has meant, among other things, using common standards to integrate the business data owned by each of the DoD Services Components. The mindset that creates twenty year weapon system development cycles plans for the "next major war". But this long-term mindset is unresponsive to current war fighter needs in today's battlefield and theater.

## Governance Structure

The Department is employing industry and warfighter best practices in designing and managing its transformation effort. DoD has adopted a governance structure that implements a tiered accountability principle based on the existing organizational structure of the Department. These tiers include a "corporate" or DoD Enterprise level and a Component level (the Military Departments, Defense Agencies, DoD Field Activities, and Combatant Commands). This distinction recognizes that, while the Secretary of Defense sets the tone from the top, each of the Components has its own way of achieving its mission, its own natural constituencies, and its own appropriations.

Component-level business transformation is the responsibility of the Component headquarters. Components develop strategies, schedules, and budgets for their Component transformation, then implement these plans. Components are responsible not only for executing their individually assigned missions, but also for ensuring that joint operations run smoothly, and that information flows freely across the enterprise so DoD can function as a cohesive whole.

An enterprise-level governance structure for the information enterprise includes the need for data standards, business rules, specific systems, and an associated integration layer of interfaces for the Components. These standards need to be established through joint cooperation, and they represent the rules of engagement for all of the DoD Components. The idea is that each Component's transformational program increases the DoD's ability to reap the benefits of improved information exchange across organizational boundaries. The expectation is to drive the DoD down the path to interoperability and accelerate the Armed Force Services' transformation efforts.

## End-Goal Objectives

The end-goal objectives are to allow the DoD business model to address the threats of terrorism, counteract military challenges throughout the globe, and keep pace with the changing nature warfighting. This type of transformation is about instantaneously reacting to the demands of today's modular, agile, and technologically advanced joint forces. This new reality quickens the pace of business operations. It is no longer just a matter of achieving cost, schedule, and performance requirements; it is also about spawning business agility. The objectives for the future can be summarized to:

- Support the joint warfighting capability of the DoD
- Enable rapid access to information for strategic decisions
- Reduce the cost of defense business operations
- Improve financial stewardship to the American people

## Assumptions for the Information Enterprise

Based on the transformation goals, we can now formulate the assumptions necessary to direct the information enterprise towards its target state. The current DoD follows a hierarchical organization:

- The Office of the Secretary of Defense sets the overall strategy, provides oversight, and manages military and defense capabilities.
- The vast organizations underneath are known as DoD Component organizations and include the Armed Forces, and DoD business operations that includes finance, logistics, human resources and medical commands.
- All of these component organizations have their own way of doing business and their own set of users.

The objective of transforming the entire DoD information enterprise is to ensure that DoD headquarters continue to provide guidance and communications. And each Component organization is able to execute their assigned missions, conduct operations in a joint manner and ensure that information is shared in an easier manner than before.

To facilitate the transformation process toward net-centric operations, DoD has developed a set of core assumptions that helps to move forward to the target state:

- **The Concept of Federation** – Within a large enterprise, it is expected that there are a significant number of component organizations. The federation concept allows each sub-organization to work in an autonomous manner. However, federation ensures that decision makers and program managers are able to align their programs and capabilities across the various tiers of the larger enterprise.
- **The Concept of Tiered Accountability** – Within the DoD, tiered accountability aligns the responsibility for decision making and execution through each of the DoD Component organizations. For example, this allows accountability for a joint mission to be coordinated with the Army, Navy and Air Force, along with the joint commands.

The transformation of the information enterprise ensures that each tier within the organization governs the areas for which it is responsible, is able to acknowledge and maintain consistency with the guidance that is provided by the higher tiers. The federated decision making process allows each tier to leverage the decisions and services of the other tiers.

## Net-Centric Assumptions

Obviously, the ability to achieve a rich information enterprise demands a cultural shift regarding how information is considered, how it is shared and how it is acted upon. If the information is only stored within a particular organization's own repository, the information remains hidden and presents no use to other organization members who may have use for it. The ability to move to a net-centric mode allows the information enterprise to meet the needs of unanticipated users who require the information to be visible and shared. Sufficient context needs to be available to understand the data and services that are available and to determine suitability for a particular purpose.

The new target state that follows net-centric principles can scope out data and services that exist and what can be made visible and accessible. As a result, the concept of information stovepipes would be eliminated and decision-making agility and speed would be increased. Regardless of time or place, DoD organization users should be able to say that they get the information that they need to perform their mission.

## DEFINING THE DEFENSE INFORMATION ENTERPRISE

The need for a Defense Information Enterprise is to provide a common foundation to support accelerated transformation of the DoD to net-centric operations. The expectation is for the information enterprise in the long term to achieve the net-centric vision. Here is an excerpt of the definition of the Defense Information Enterprise, as provided within the Defense Information Enterprise Architecture v1.0 document published in April, 2008.

The Defense Information Enterprise (DIEA v1.0, 2008) consists of:

- All information resources, assets, and processes required to achieve an information advantage and share information across the DoD and its mission partners.

In specific, the enterprise includes:

a. The information itself, and the DoD's management over the information life cycle;
b. The processes, including risk management, associated with managing information to accomplish the DoD mission and functions;
c. Activities related to designing, building, populating, acquiring, managing, operating, protecting and defending the information enterprise;
d. Related information resources such as personnel, funds, equipment, and information technology, including national security systems.

### Principles of the Defense Information Enterprise

The following are global principles that have been excerpted from the Defense Information Enterprise Architecture (DIEA v1.0, 2008) that broadly reflects the intent of the target state of the Defense Information Enterprise. Net-centric goals and duties will need to adhere to these organizational principles.

- DoD CIO-governed information resources are conceived, designed, operated and managed to address the mission needs of the DoD.
- Interoperability of developed solutions across the DoD is a strategic goal. Solutions should fit into all parts of the GIG and must work together to achieve this goal.
- Information within the enterprise is made interoperable by following the rules for net-centric sharing of data and services across the enterprise.
- The objective is to gain infrastructure interoperability through definition and enforcement of standards and interface profiles and implementation guidance.
- Data assets, services, and applications on the GIG shall be visible, accessible, understandable, and trusted to authorized (including unanticipated) users.

*Figure 1. Reference Model for the DoD Information Enterprise*

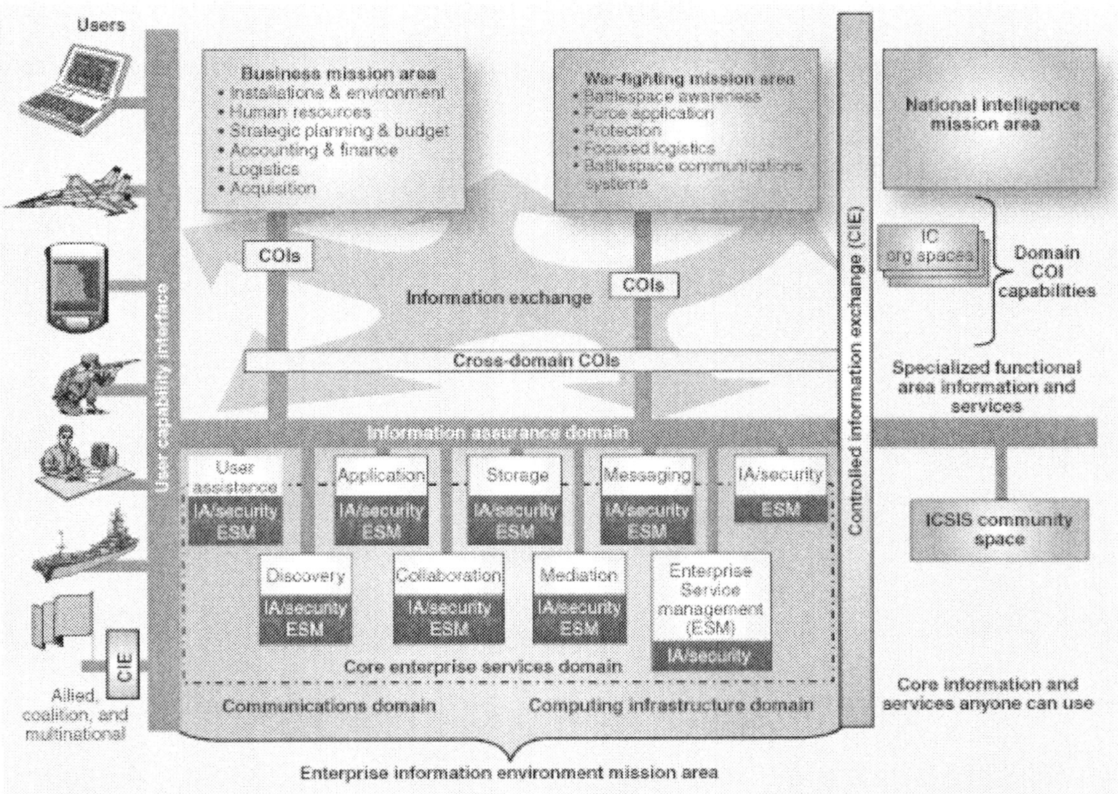


- Defense Information Enterprise services shall advertise service-level agreements (SLAs) that document their performance, and shall be operated to meet that agreement.
- The GIG will provide a secure environment for collaborative sharing of information assets (information, services, and policies) with DoD's external partners, including other Federal Departments and Communities of Interest (e.g., Department of Homeland Security, the Intelligence Community), state and local governments, allied, coalition, non-governmental organizations (NGOs), academic, research and business partners.

## REFERENCE MODEL FOR DoD INFORMATION ENTERPRISE

As part of the defense transformation toward Net Centricity, a reference model is shown below in Figure 1 that represents a target state for the information enterprise. It is obvious that this may not be the end state that truly describes the net-centric vision.

However, this reference model is currently in use within the DoD (DoD EA Transition Strategy, 2008) and delineates the federated responsibility for each of the DoD functional areas, and provides tiered accountability for military and government decision making.

Figure 1 certainly shows the complexity of the entire DoD and military organizations, and works to make it into a services-based community. As you analyze the diagram from left to right, all of the different user communities have been identified based on example icons. The diagram then provides an understanding as to how information is compartmentalized based on critical functions and duties. It also identifies the information infrastructure support that is necessary.

The Defense Information Enterprise reference model thus illustrates a cohesive mechanism to promote net-centric goals and vision that enables the entire DoD and military to move ahead beyond a command and control structure.

Here is an explanation of the elements covered within the Figure 1 illustration:

## User Access

- Users are accessing information from laptop computers using, for example, the World Wide Web.
    - information as they are within a fighter jet airplane
    - mapping information from their handheld computing device
    - information as they are targeting to shoot enemy in the battlefield
    - information as they are sitting at their desk in administrative duties
    - information inside a naval warship as it is sailing across the waters
    - accessing military allied, coalition and multinational forces information

## User Capability Interface

- The reference model addresses the need for a common interface and a common set of tools to be provided to the vast set of users within the enterprise
- Each set of users have to be able to view a full set of information that needs to be available to all parties
- Users also need to be able to access information based on their mission area and their "need to know" based on information assurance guidelines

## Enterprise Mission Areas

- Business Mission Area includes the following DoD business functions as listed within the diagram:
    - Installations and Environment
    - Human Resources
    - Strategic Planning & Budget
    - Accounting & Finance
    - Logistics
    - Acquisition
- War Fighting Mission Area includes the following DoD business functions as listed
    - Battlespace Awareness
    - Force Application
    - Protection

*Figure 2. Net-Centric Goals for Defense Information Enterprise*

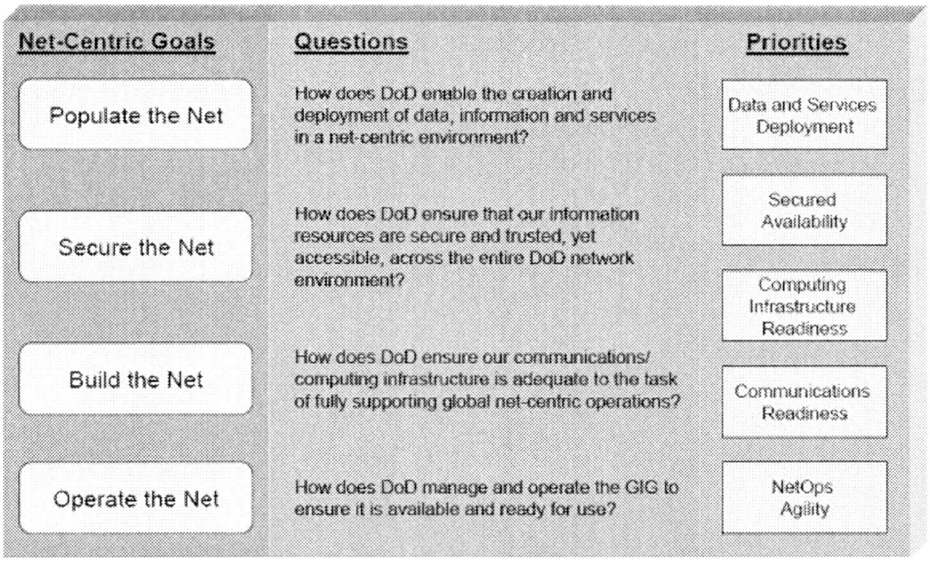


- Focused Logistics
- Battlespace Communications Systems
- Enterprise Information Environment Mission Area (EIEMA)
  - This mission area provides the information infrastructure to support the other key areas of business and war fighting missions
  - The information infrastructure into the following domains:
    - Communications Domain
    - Computing Domain
    - Core Enterprise Services Domain
    - Information Assurance Domain
- National Intelligence Mission Area includes all of the military intelligence activities for gathering Intelligence, Surveillance and Reconnaissance. It includes:
  - Intelligence Community organization spaces
  - Specialized information area functions and services that serves secure, compartmentalized information

## Controlled Information Exchange (CIE)

- The illustration shows that to address the needs of the intelligence community, information sharing needs to be filtered and controlled based on a CIE
- Based on DoD Information Assurance guidelines, NIPRNet and SIPRNet networks and communications have different set of protocols for communication
- Information exchanges to the intelligence community require high information assurance guards to protect information sources, access and sharing

## Communities of Interest (COIs)

- DoD transformation capabilities are based on the use of Communities of Interest
- Each mission area and mission segment or domain have their set of COIs
- There are also Cross-Domain COIs that go across multiple mission areas
- COIs provide the critical functions of information sharing and information exchange throughout the Defense Information Enterprise

## Core Enterprise Services

The EIEMA has developed a set of core enterprise services that address user needs and takes care of the infrastructure based on a Service Oriented Architecture methodology. The core enterprise services constitute the following information elements:

- User Assistance
- Application
- Storage
- Messaging
- Information Assurance/Security
- Discovery
- Collaboration
- Mediation
- Enterprise Service Management

# INFORMATION ENTERPRISE GOALS FOR NET CENTRICITY

The net-centric goals (DIEAv1.0, 2008) can be simply put as provided below within Figure 2. It addresses the overall goals for the entire DoD and military to create a target information enterprise as necessary for net-centric force transformation.

The information within Figure 2 flows according to the goals addressed, questions raised, and priorities created.

- **Goal 1: Populate the Net**
  - How does DoD enable the creation and deployment of data, information and services in a net-centric environment
  - This leads to the priority for Data and Services Deployment
- **Goal 2: Secure the Net**
  - How does DoD ensure that our information resources are secure and trusted, yet accessible across the entire DoD network environment?
  - This leads to the priority for Secured Availability
- **Goal 3: Build the Net**
  - How does DoD ensure that the communications and computing infrastructure is adequate to the task of fully supporting global net-centric operations?

  - This leads to the priority of Computing Infrastructure Readiness
- **Goal 4: Operate the Net**
  - How does DoD manage and operate the Global Information Grid to ensure it is available and ready for use?

## DEFENSE INFORMATION ENTERPRISE STRATEGY DETAILS

This section provides the details on the DoD strategy for the Defense Information Enterprise. The net-centric vision assumes a rich information sharing environment in which data and services is widely available, easily discoverable, usable and trusted across the GIG. There needs to be sufficient context available to understand the data and services that are available and to determine suitability for a particular purpose. The ultimate goal is to have all data and services to be visible and accessible. In the new context, users should be able to say, "I can get the information I need to perform my mission."

For the Data and Services Deployment (DSD) priority, the DoD intent to achieve the net-centric vision, is to transform its approach from deployment of systems to the delivery of information and services and abides by definitions, rules and principles.

### Enabling the Data and Services Environment

To effectively provide services and support to information providers and consumers, a number of near-term issues need to be evaluated:

- **Enable the Practice of Service Orientation** – As new capabilities are defined, system solutions should be considered that encourage a service-oriented approach within the net-centric environment. The enterprise needs to be trained such that there is a common understanding of key service-oriented architecture concepts. This includes separation of interfaces from implementations, and separation of business logic from infrastructure functions.
- **Initiate Communities of Interest (COIs)** – The COI approach allows large disparate organizations to collaborate on high priority data, information and services issues. These collaborative communities can address information sharing gaps by identifying the most important data and capabilities needed to support agile and collaborative community business processes.
- **Enable Information Discovery** – We want to make sure that users are able to easily find data and services in the net-centric environment. Users should be able to obtain services from authorized sources, and the information needs to be tagged with metadata at the time of creation, not retroactively. Information discovery requires content discovery brokers that scan information and service registries across the GIG to locate requested information content.
- **Create Service Contracts for External Interfaces** – Service contracts need to be written so that services are discoverable, understandable, and usable. This requires information providers to register their services and provide details that allow consumers to use, manipulate or transform data.
- **Support Shared Business Models for Service Operation and Sustainment** – Instead of promoting stove-piped funding strategies for development and sustainment of systems, the new model is to provide end-to-end solutions for applications, data and underlying hardware. The idea is to

accommodate shared expenses between information providers and consumers.

- **Establish Oversight and Governance for Common Services** – The enterprise needs to have a set of essential core services for all common functionality and primary interfaces. These services require technical, operational, and programmatic oversight and governance on an ongoing basis.

## Secured Availability

It is a well-known fact that DoD networks and information are constantly threatened by a variety of adversaries, including terrorists and criminal organizations, rogue nations, insiders and common hackers. Since most DoD systems are meant for decision making, the GIG support infrastructure will always be a high priority target. Secured availability provides the reliability, and resiliency of the GIG infrastructure to successfully maintain information superiority. Secured availability needs to also work within the principles of the current DoD Information Assurance Certification and Accreditation Process.

Secured availability addresses several challenges DoD faces in achieving a fully net-centric environment. This includes ways to protect and secure critical data and capabilities, while providing authentication and non-repudiation of GIG information and transactions. It also makes it possible to rapidly and securely respond to incidents threatening GIG operations. It is understood that secured availability in a net-centric environment requires new technologies, new policies and new levels of collaboration. This would extend from the DoD to its federal, state, local, industry and coalition partners. Key elements of successful implementations include:

- Assessing threats and risks associated with the software, hardware and services supply chain to enable DoD program, security, and operations personnel to understand the level of trust that can be associated with the IT components they acquire, manage or use.
- Providing and managing assured identities for all users, services, and devices to facilitate dynamic information sharing within and across the network boundaries of organizations at varying trust levels.
- Permanently binding metadata to associated data objects at the time of the object's creation, to facilitate assured data visibility and handling.
- Enabling rapid modification of access, resource allocation and prioritization of bandwidth, processing, and storage through enterprise-wide, policy-based management.
- Improve the management of enterprise-wide security services and infrastructure to handle encryption, crypto key management, identity privileges and security audits.

## Computing Infrastructure Readiness

The rest of the DoD priorities that align to net-centric goals deal with DoD's shared infrastructure environment. These include Computing Infrastructure Readiness, Communications Readiness and NetOps Agility. As part of the GIG infrastructure, computing infrastructure readiness accounts for the following shared infrastructure principles:

- GIG enables connectivity to all authorized users
- GIG infrastructure capabilities need to be survivable, resilient, redundant, and reliable to enable

continuity of operations and disaster recovery

- This may be in the presence of attack, failure, accident, and during natural or man-made disasters.
- GIG infrastructure needs to be scalable, changeable, deployable and manageable of unexpected users.

The computing infrastructure needs to take into account users at the forward or leading edge of the mission operations environment. This means support of end-users operating in environments challenged by intermittency and low bandwidth. Characteristics of this new computing infrastructure include use of:

- **Virtualized application environments** – This means that applications are hosted in shared environments that allow dynamic changes to processor and storage capabilities depending upon usage patterns. This virtualized environment needs to provide seamless access to all applications and services regardless of their physical location.
- **Location-independent storage** – This allows distributed storage architecture that allows services and applications will share storage located anywhere, and allows consolidation and efficient use of data storage resources.
- **Dynamic, automated storage provisioning** – This means that knowledge and usage patterns will be used to heuristically allocate data storage. This dynamic provisioning is based on devices to be able to learn from past usage experience to better serve users.
- **Automated status reporting** – The idea is that all GIG computing infrastructure resources can continually report the status of their assets, thus enabling network operations to have a continuous view of the status of all resources.

## Communications Readiness

Communications readiness is the next step in providing the shared infrastructure services that is a key to providing the net-centric operational environment with a dependable, reliable, ubiquitous network. This type of readiness ensures that effective information transport is provided for the computing infrastructure, intelligence and network operations priorities.

Changes to the information transport environment require long-term planning and implementation. A full-scale change to the environment needs to be codified in policy, procedure and guidance. Communications and transport planning includes evaluation of the quality of equipment and technology, capacity planning and preparing for redundancy.

At this time, the enterprise strategy for providing communications and transport includes:

- **Modularization** – Communication and transport solutions need to be modular, IP-based, and will be based on historical usage patterns, location and mission need. Acquisition of solutions can use a standard hardware bill of material that streamlines the acquisition process and results in faster deployment or enhancement. The idea is to also reduce training requirements and promote confidence in the overall transport architecture.
- **Technology Evolution** – For upcoming communications technology, the idea would

be to support Internet Protocol version 6 (Ipv6), network management Simple Network Management Protocol version 3 (SNMPv3) and the latest capacity planning modeling and simulation technology. Periodic technology refreshment needs to be inserted as part of the technology evolution process. This would lead to strategies such as establishing tiered network security, set up of newer services and federation of existing services.

- **Limiting Uniqueness** – Based on the adoption of a standard bill of material, the idea would be to eliminate non-standard equipment and their associated support requirements. This will then limit uniqueness and facilitate an interoperable network that reduces spare parts inventory, hardware and software changes and repair costs, and optimize equipment reuse.
- **Rapid Deployment** – The logistics and planning efforts would be reduced that allows for rapid deployment of communications infrastructure across the globe. Planning for the effort requires adhering to GIG capabilities in response to new mission requirements and tactical needs.

## NetOps Agility

The vision of NetOps is to transform existing and new capabilities into a force multiplier so that it enables DoD to fully employ the power of the GIG. The corresponding mission is then to employ a unified, agile, and adaptive GIG that is:

- **Mission Oriented** – All information-dependent processes necessary for a mission can be effectively supported
- **User Focused** – Each user can access and obtain needed information from anywhere in the GIG in a timely manner; even when their needs are unanticipated
- **Globally Agile** – Rapidly changing mission priorities can be met by dynamically maneuvering GIG resources

Like much of the GIG, NetOps today is delivered through organizational and functional stovepipes with varying degrees of interoperability and information access. Each of these stovepipes has its own, largely independent management capability, which seldom shares information regarding the status of its management domain. The Joint NetOps Concept of Operations has enabled the DoD to begin significantly improving how the GIG is operated and defended. For NetOps to effectively play its role in enabling net-centric operations, however, major challenges will have to be addressed:

- GIG Situational Awareness information must be available to Commanders
- GIG Command and Control capabilities must support rapid decision making
- NetOps operational policies must be clear and well integrated
- NetOps must address the use of the electromagnetic spectrum
- Standardized metrics must measure the health and mission readiness of the GIG
- Capability development must be centrally governed
- Greater coordination or synchronization is required among the many independent NetOps acquisition and fielding activities currently under way

Addressing these challenges will significantly improve the ability of the operators and defenders of the GIG to fully support ongoing war fighting and peacekeeping missions in an increasingly joint and multi-partner environment.

## REVIEW OF CHAPTER GOALS

The goals of this chapter were to address:

1. Where the DoD and the military heading with regard to a target state of the information enterprise:
   - Provide a perspective on how to manage tomorrow's information
2. What is the strategy for transforming the DoD to a target state:
   - Provide the vision and mission of the national defense strategy
3. Define the details of a defense information enterprise:
   - Provide a full definition of the enterprise and provide a set of global net-centric principles
4. How we can pictorially depict the elements of the defense information enterprise:
   - Discuss each of the organizational element of the overall enterprise
5. The net-centric goals and priorities for the target enterprise:
   - Provide the detailed goals and the set of named priorities
6. The details of each of the DoD priorities for the enterprise:
   - Discuss the aspects of data and services deployment, secured availability, communications infrastructure readiness, communications readiness and NetOps agility

## REFERENCES

Department of Defense (2007, September). *Enterprise Transition Plan (ETP), Defense Business Transformation Overview*. Washington, DC: Author.

Office of the Assistant Secretary of Defense for Networks and Information Integration, Department of Defense Chief Information Officer. (2008, April). *Defense Information Enterprise Architecture,* (Version 1.0).

Office of the Assistant Secretary of Defense for Networks and Information Integration, Department of Defense Chief Information Officer (2008, February). *Enterprise Architecture Transition Strategy,* (Version 2.0). Department of Defense Business Transformation (2008, September). *Enterprise Transition Plan, Appendix D.*

Office of the Assistant Secretary of Defense for Networks and Information Integration, Department of Defense Chief Information Officer (2007). *Department of Defense Information Sharing Strategy* (Version 1.0). Bass, T., Mabry, R., (2004, April). *Enterprise Architecture Reference Models: A Shared Vision for Service-Oriented Architectures*, (Version 0.81). IEEE Milcom 2004.

Office of the Secretary of Defense. (2005). *The National Defense Strategy of the United States of America.* Retrieved from www.defenselink.mil

# Chapter 6
# Net-Centric Military to Civilian Transformation

## CHAPTER CONTENT

As you explore through Chapter 6, it covers the following topics:

- A Look back at Command & Control Principles
- Evolution of Net-Centric Principles from Command and Control
- Industry Shift toward Net-Centric Systems
- DoD to Civilian Technology Transfer
- Industry Topics related to Net Centricity
- Upcoming Industry Technology Areas
- Transformation Perspective on Net-Centric Computing
- Perspective from an industry veteran on the effects on today's industry

DOI: 10.4018/978-1-60566-854-3.ch006

## CHAPTER FOCUS

This chapter focuses broadly on the technology effects of the net-centric transformation on our industry and civilian market. We start by discussing aspects of command and control principles, how it has run its course within the military and how the hierarchical nature is changing to embrace a matrixed decision making process. We then discuss the evolution of net-centric principles and how the government contracting and civilian industry is being affected by this transformation. We then discuss specific details on DoD to civilian technology transfer. A broad set of industry topics are then discussed and is categorized based on net-centric needs. Then we provide a glimpse into upcoming industry technology areas that have shown promise. We end with a perspective from a guest author, who provides an industry perspective on the use of net-centric concepts and proliferation of net-centric computing.

This chapter continues on our transformation perspectives, and steps through the net-centric future from a system perspective. We first review the current nature of the DoD that has worked based on a command and control structure. This then is giving way to the net-centric mechanisms for systems operations.

We examine the net-centric transformation from the military to the civilian community and commercial marketplace. It is important to look at defense to commercial technology transfer principles. Then examine how current command and control systems run by government contractors are shifting their focus to net-centric principles. It is also important to figure out how commercial entities are taking advantage of net-centric concepts within the computer and communications industry. There has been an emergence of "Net-Centric Computing" as a concept that we further explore. The types of technologies that are involved in the computing and communications infrastructure are discussed. We then provide a guest author perspective as an industry subject matter expert on the net-centric transformation and industry initiatives.

## A LOOK BACK AT COMMAND AND CONTROL PRINCIPLES

Our defense infrastructure has been relying on a complex set of command and control principles for ensuring that effective operation takes place for military forces. For the past few decades, a number of command and control principles have been implemented within technology systems that are currently in use by Army, Navy, Marines and Air Force for day to day operations. The ability to command and control has allowed the military to hierarchically set up their organizational structure to make decisions in an orderly manner.

Command and control systems have been responsible for the gathering, processing, disseminating, and displaying information that is needed for execution of the Armed Forces. Typical characteristics of a command and control system have been the following:

- Requires the active need of human operators who provide real-time command decision-making to make each command and control system work
- Requires systems to be custom-created and deployed to the field across the globe
- Command and control decisions are made in a serial manner that starts from the top administrative unit and works its way down to field operations

If we take a look at a definition of the command and control principle, one can be provided by DoD Joint Staff, (Joint Publication 1-02, 2001) as the following:

**Command and Control:** The exercise of authority and direction by a properly designated commander over assigned and attached forces on the accomplishment of the mission. Command and Control functions are performed through an arrangement of personnel, equipment, communications, facilities and procedures employed by a commander in planning, directing, coordinating and controlling forces and operations in the accomplishment of the mission.

From a technology system perspective, command and control systems allow the designated military commander the ability to readily exercise his or her authority and direction over their organization. Command and control systems are known for their real-time capabilities and offer multiple set of functions to address ongoing operations. These include the following:

- Strict control and hierarchical management of personnel at each level of a military organization
- Arrangement of equipment, facilities, and communications based on decisions of the top-level command and the allocation of resources based on command and control authority
- Accomplishment of mission and system objectives based on a hierarchical information flow on a "need to know" basis
- Operational and field activities conducted based on strict enforcement of the chain of command and personnel assignments

This set of command and control principles have served the military well over the past century. However, as technology has progressed and information capabilities have increased by leaps and bounds, strict adherence to command and control principles have become antiquated. Command and control principles do have give enough freedom for decision making to be done in the field.

Agile decision making requires a matrix-based decision making framework that allows inputs from a variety of active stakeholders. Command and control systems may perform very well on a real-time basis, but they may not provide the types of information that personnel at the field level require or desire. Other detractions of command and control system are that they lack the ability to be easily configured, do not allow for automated control processes are that they do not evolve easily based on technological upgrades, and do not allow for reuse by multiple users, or in multiple environments.

## EVOLUTION OF NET-CENTRIC PRINCIPLES FROM COMMAND AND CONTROL

The DoD has been working on development approaches to modernize their command and control systems to net-centric principles as part of Network Centric Warfare. These include the following main tenets for future systems:

- Continuous operational evolution to new types of warfare
- All-encompassing user connectivity to distributed and authoritative data
- Dynamic growth in command and control capability
- Incremental change to systems based on fast and agile implementation

To allow command and control systems to keep their desired properties and yet transform to the new net-centric paradigm, it would be best to address the following system functions:

- **Configurability** – Components within a command and control system usually vary depending on the particulars of each given military situation. So the idea is to assemble new net-centric systems in different configurations, which address system configurations that have not been anticipated in advance.
- **Interoperability** – The components of a command and control system must share information with another system in a mutually understandable and meaningful manner. New net-centric systems would include components that were not initially designed with interoperation in mind, and the specifications were also not explicitly written.
- **Evolvability** – To accommodate technological advances, net-centric systems would allow possible changes to be made to specific components of the system without extensive modifications done to other components of the system.
- **Robustness** – A new net-centric system must behave in a reliable manner in normal operation. The system will often be used in situations where its elements are subject to temporary or permanent failure. This includes temporary loss or overload of communications links, and destruction of processing elements. A new net-centric system must fail gracefully and recover automatically in most situations.
- **Security** – A new net-centric system must protect its information from unauthorized disclosure and alteration, and must maintain the user's access to this information. This includes all of the security principles such as availability, integrity, authentication, confidentiality, and non-repudiation.

These operational system characteristics need to be part of the contract requirements for building new net-centric systems. The government is currently working with commercial vendors and contractors to ensure that these operational characteristics are adhered to, by new systems that are built.

## INDUSTRY SHIFT TOWARD NET-CENTRIC SYSTEMS

As the military and associated government entities transform their organizations, it allows for a set of large-scale and possibly disruptive changes. Industry leaders such as large government contractors, research organizations, and small entrepreneurial companies have been pondering about the march toward net-centric systems. Commercial strategies to capitalize on new net-centric capabilities, however, have not yet taken hold.

The commercial marketplace that works in hand-in-hand with the government and military, currently develop programs for civilian applications, integrated communication systems, networked sensor systems, and command and control systems. Industry leaders do believe that a rapid shift in military services will occur from platform-based systems that operate independently to a network-centric integrated computing infrastructure. This new infrastructure is necessary to operate tomorrow's planes, tanks, ships, and space-based weapons.

The commercial hardware and software industry that cater to the military realize that today's systems leave a lot to be desired. An effective integration mechanism is yet to be developed to take into account a number of the information challenges such as security, availability, interoperability, and large system costs.

The complaint by a number of government contractors is that most current approaches to net-centric software has been to try to "patch" the current systems that are in place. The reality is that patching together a set of disparate systems actually increases the difficulties in interoperability and thereby makes the systems less net-centric. The long-term goals for the hardware and software industry would be to upgrade the infrastructure from the bottom-up. Net-centric software may have fixed functions but they require that software services are created dynamically and then verified and validated real-time, without the necessity of human intervention.

### Industry Roadmap towards Net-Centric Systems

The government and civilian contractor marketplace is starting to develop industry roadmaps toward the following upcoming technologies:

- A net-centric semantic infrastructure
- Sensory-based systems
- Knowledge management
- Integrated cognitive systems

Industry direction would include a variety of market trends that include a detailed examination of technologies, standards, end customers, suppliers, industry sectors, software and hardware applications, and early adopters. Based on this information, industry leaders and company executives would have to plan for new products, services and solutions. A new direction toward net-centric systems would try to solve the issues of complexity, scale, and costs – leading to future breakthroughs in technology and product development.

## DoD TO CIVILIAN TECHNOLOGY TRANSFER

The DoD has been promoting technology transfer of military technology to the civilian sector for many years. Agencies such as DARPA (Defense Advanced Research Project Agency) and other eminent organizations have been trying to foster research and development such that U.S. citizens can broadly benefit from advanced technology developed by the military.

Industry-led activities that include a coalition of government and private organizations realize that Net Centricity and its ensuing set of distributed technologies pose the same set of structural challenges. It is important that when technology strategies have commercial implications that these should be adopted by the technology industry in general.

### DoD Domestic Transfer Policy

It is DoD policy to use domestic technology transfer as an integral part of the research and development (R&D) effort based on a variety of technology transfer mechanisms. This encourages technology transfer from all appropriate R&D activities, consistent with the military mission. As part of DoD policy, the commanders and directors of Army, Navy, or Air Force laboratories and centers have the responsibility and the authority to enter into cooperative research and development agreements (CRADAs) in accordance with federal law. This allows the government to be able to license, assign, or waive rights to intellectual property developed by the joint military-civilian activity. This also allows the government to support active marketing and assistance by their laboratories or centers to interested civilian organizations and corporate entities. The DoD works with private companies based on participating in economic development organizations, contracting with partnership intermediaries, and providing technical assistance to State and local governments and local educational systems.

### Cooperative Research and Development Agreements

Cooperative Research and Development Agreements or CRADAs are set up by government organizations to conduct research and development (R&D). These R&D activities are usually for a specified duration and are targeted to be consistent with the DoD mission at hand. Special consideration is given to entering into CRADAs with small-business firms located in the United States and industry consortia involving small-business firms. The expectation is that businesses who develop new products and inventions under the CRADA umbrella, or produce and manufacture the goods substantially in the United States. R&D efforts from a CRADA partnership may end up in future government procurement activities, where knowledge is then transferred to government and military acquisitions that foster the latest technology.

### DoD Laboratories and Research Centers

The DoD and the defense department have a large number of laboratories and research centers who promote the concepts of DoD to civilian technology transfers. Commanders and directors of laboratories and centers ensure that U.S. domestic businesses are directly involved in the development and roll-out of new technologies.

There are organizations within our government that are known as an FFRDC. This moniker stands for a federally funded research and development center that are unique independent nonprofit entities. Examples of FFRDC for the defense department include the MITRE organization, MIT's Lincoln Labs, Battelle Institute and others. These are sponsored and funded by the government to meet specific long-term technical needs that are not regularly pursued by commercial entities. The intent of FFRDCs is to assist government agencies with scientific research and analysis, systems development, and systems acquisition. They work both with government organizations and commercial entities to bring together the confluence of government, industry, and academia to solve complex technical problems.

## INDUSTRY TOPICS RELATED TO NET CENTRICITY

This section provides a list of current industry topics that is related to the complex concept of Net Centricity. This is by no means an exhaustive list. However, it is instructive to provide this information in this section as part of the transformation perspective. Most of these topics are being covered within this book in different chapters.

It is important to highlight that small and large businesses have been working on these topics for the past several years that include various aspects of the net-centric infrastructure. As contractors help the government and military, it is best to highlight the set of technology topics that extend the net-centric enterprise to corporate environments. The following steps through these topics in a categorical manner.

Category: Networking and Communications

- **Network Architecture** – current Internet topology, layered and hierarchical structure of a communications network
- **Use of Networked Applications** – use of e-mail, FTP, wikis, web browsers, instant messaging
- **Networking Protocols** – use of HTTP, TCP/IP, UDP, remote procedure calls, DHCP, and other common protocols
- **Distributed Computing** – client/server architecture, peer to peer, hub and spoke
- **Internetworking and Routing** – routing algorithms, internetworking, congestion control
- **Transport Layer Services** – connection establishment, performance issues, flow and error control
- **Physical and Data Link Layer Concepts** – includes framing, error control, and flow control protocols

Category: Web Technologies

- **Web-based programming** – Server-side programs, client-side scripts, the applet concept
- **Web Server Configuration** - Handling permissions, file management, custom network configuration, use of Web protocols, Web-based transmission
- **Common Server architectures** – Role of client computers, business rules server, nature of N-tier architecture
- **Web Content Management** – use of Internet information servers, web site creation and management, publishing information and applications
- **Grid Computing** – web-based storage, distributed grid storage at different hubs, use of web-based clusters, or mesh
- **Web 2.0 Software** – use of dynamic front-end applications, AJAX, customer relationship management over the web, enterprise content management
- **Web Interfaces** – Browsers and APIs, Web Search technologies, database-driven web sites, remote procedure calls
- **Web Services** – use of Service Oriented Architecture, web-based middleware, distributed object systems, lightweight directory services

Category: Network Security

- **Use of Cryptography** – Secret-key algorithms, Public-key algorithms, Authentication protocols, Digital signatures, Kerberos
- **Network Attacks** – Denial of service, flooding, sniffing and traffic redirection, message integrity attacks, identity hijacking, exploit attacks such as buffer overruns, Trojan horses, backdoors, inside attacks, infrastructure attacks, DNS hijacking, router misbehaving by dropping packets
- **Authentication** – login procedures, use of passwords and access control mechanisms

- **Firewall Protection** – intrusion detection services, network defense tools, detection of malware, prevention of un-trusted packets
- **Network Management** – resource management, auditing and logging, network address translation, virtual private networks
- **Use of IPSec** – stronger set of Internet security protocols, adhere to new Ipv6 standards

Category: Multimedia Technologies

- **Multimedia Standards** – audio, music, graphics, image, telephony, video, TV
- **Multimedia Technologies** – sound and audio, image and graphics, animation and video, digital data
- **Input and Output Devices** – scanners, digital camera, touch-screens, voice-activated, MIDI keyboards, synthesizers
- **Storage Standards** – Magneto Optical disk, CD-ROM, DVD, multimedia servers and file systems
- **Analog and Digital Representations** – Encoding and decoding algorithms, sampling and quantization for digital representation, lossless and lossy compression, data libraries
- **Performance issues** – timing, compression factor, suitability for real-time use
- **Compression Technology** – Data compression for example Huffman coding and the Ziv-Lempel algorithm, audio compression and decompression, image compression and decompression, video compression and decompression

Category: Mobile Computing

- **Evolution of Wireless Standards** – Upcoming IPv6 standards, wireless telecommunications standards, wireless transmission standards
- **Wireless Networks** – Wireless local area networks and satellite-based networks, Wireless local loops, cell phone networks, point to point network
- **Mobile IP protocol** – differences between IP protocol with regard to mobility management, location management, and performance
- **Mobile data access** - Server data dissemination, client cache management, mobile aware adaption
- **Software Support for Mobile and Wireless Computing** - Extending the client-server model to accommodate mobility, role of middleware and support tools, emerging technologies

## UPCOMING INDUSTRY TECHNOLOGY AREAS

Now that we have provided a glimpse of the types of technologies that are being pursued by the technology industry, it would be instructive to discuss in more detail a set of new and upcoming technologies. These technologies are an outgrowth of the military transformation to a net-centric future. However, these technologies are being pursued by the commercial marketplace primarily for civilian and consumer purposes.

## Semantic Technology and Infrastructure

The use of semantic technology and the ensuing business infrastructure should be viewed as a cohesive consolidation of the semantics of different business areas. This is intended to allow the alignment of semantics between different organizational units within an enterprise. If a business or military organization adds a semantic infrastructure, this provides the necessary facilities to enable:

- The exchange of a common language for exchanging business messages
- The ability to support intelligent routing of data over an enterprise service layer
- An intelligent model driven data transformation

In the next decade, organizations will deploy a semantic infrastructure to achieve a low latency and low impedance solution to business challenges. The direct benefits are that there will be an increased transparency in business performance metrics and business activity monitoring. Key components of a semantic infrastructure that would deploy to commercial or military enterprises would be the following:

- An Enterprise-level ontology
- A significance language for the enterprise
- An intelligent data delivery platform
- An intelligent data transformation framework

An enterprise level ontology helps to define data conventions, metadata, data structure, and an information glossary for the enterprise. It represents the infrastructure components within the enterprise architecture and is coordinated with the organization's business models.

Another key component to the semantic infrastructure is a well designed business significance language. Such a language is a prime vehicle that can be used by the information services architecture of an enterprise. The business significance language should be decoupled from the enterprise ontology, should be able to facilitate discourse on all business topics, and the language should be free of all topical semantics such as character encoding.

Semantic technology offers an intelligent data delivery platform that allows the semantics of messages to be exploited for routing purposes. This means that the message is delivered without compromising the integrity, confidentiality or authenticity of the message payload. The business significance language facilitates the exploitation of the semantics defined in the enterprise ontology and helps towards the sustained improvement of the data delivery service.

A data transformation framework is also necessary for the enterprise since all applications and data sources will not be able to deliver data with the structure and semantics required by the business significance language. Hence, a set of semantic mappings and transformations will be needed to accommodate communications with these applications. A data transformation framework maps incoming and outgoing data to and from semantically equivalent data blocks.

For the DoD and military, semantic technology is crucial in allowing secure communications to readily occur between disparate sources. A common ontology and significance language allows different communities of interest to effectively communicate within a far-flung global network.

## The Advent of Cognitive Systems

A simple definition of cognitive systems are that they are systems "that know what they are doing". It is expected that these types of systems are in our upcoming future and will stem from the net-centric transformation. At this time, DARPA, the research agency within the DoD has been working with private contractors to develop and promote the use of integrated cognitive systems.

A cognitive system has the following properties:

- A system that can reason, by using substantial amounts of appropriately represented knowledge
- A system that can learn from its experience so that it performs better tomorrow than it did today
- A system that can explain itself and be told what to do
- A system that can be aware of its own capabilities and reflect on its own behavior
- A system that can respond in force when provoked

Cognitive systems have been the outgrowth of past research and development in neural network computing, artificial intelligence and robotics. Advances in the understanding of human neural systems allow human-level scaling of hardware technology. Cognitive technology intends to work in bits and pieces that range from searching large-scale knowledge bases to machine learning in support of data mining.

Cognitive system examples include the following modes of behavior:

- **Self-aware Software** – Extend functionality by means of interactive dialogue, actively assists programmer in the debugging process
- **Adaptive, Cognitive Networks** – The system self-reconfigures by reasoning about traffic, and evaluating anomalies, explains causes of network delays, and learns and adapts to new system attacks
- **Perceptive, Teachable Agents** – Personalize via learning, reconfigure based on natural language request, provide an answer back – what do you want me to become, perceive important threads in large amounts of data
- **Intelligent, Multi-agent Systems** – Automatic coordination with multiple components that have shared goals, allows overall cost minimization

As part of tomorrow's battlefield, it is the intent of the military to develop integrated cognitive systems that work on a set of cognitive agents. For these systems to work properly, cognitive agents will need to interact with other agents, humans, and non-cognitive systems. It is obvious that information sharing is a key component of the enterprise. Coordination and communication are essential features. But because of the autonomy and cognition that includes planning, counter-planning, and possible deceit, the system issues are much more complex than with earlier generations of computing systems.

For the commercial industry, cognitive systems may revolutionize the way business interacts with their end customers. Sensory systems that are cognitive in nature allow a business to gain a higher level of understanding of their environment and how they should capitalize on developing products and solutions that address the market forces.

## The Growth of Net-Centric Computing

Within the technology industry today, the term "net-centric computing" has been broadly defined as connected computing. It is expected that in the future, all computers will be connected in some form or another, and this connectivity would be continuous and across geographical regions. Net-centric computing isn't just communications or networking, although it certainly includes both. It is much bigger in its context and scope.

Net-centric computing covers a range of sub-specialties that includes computer communication network concepts and protocols, multimedia systems, web standards and technologies, network security, wireless and mobile computing, and distributed systems. The growth of net-centric computing mirrors the recent advances in computer and telecommunications networking, standards based on TCP/IP and the increased the importance of networking technologies in the computing discipline. The mastery of the subject area involves both theory and practice.

Net-centric computing is the commercial industry equivalent to the transformation of the military's computing and communications infrastructure. It is clear that Net Centricity is having a profound impact on computing on all levels that includes the desktop, embedded coding, servers, workstations, and the mainframe. The technology industry is bracing for the fact that the upcoming changes will affect almost every aspect of how we design and use computer hardware and software, how we design and use tools, the applications we develop, and the ways in which they will be developed.

Within a commercial enterprise, net-centric computing serves as a distributed environment where applications and data are downloaded directly from servers and exchanged with peer computing devices across a network on as as-needed basis. This is in contrast to stand-alone computers that rely primarily on local resources. Even though this sounds like net-centric computing resembles an earlier computing era several decades ago consisting of mainframes and centralized services. There are significant differences. The key to net-centric computing is that it relies on a set of virtual applications that are portable and can run on multiple platforms. This allows mobile data to be accessed via high-speed network connections, and allows low-cost computing devices or appliances to provide the local processing.

Since all computers are connected, developers who are creating software applications have a more complex problem in their hand. The developer does not have relative control of their connected environment and does not have control over all aspects of the system design. The always-connected environment forces application developers to abide by common standards that are known to all. Larger and more complex issues need to be evaluated for net-centric computing that includes system determinism, reliability, and predictability. Other challenges include problems that arise from testability and debugging in a distributed environment that is inherently non-deterministic.

Within the technology industry, net-centric computing as a computing concept will grow in its scope as the military transformation towards net-centric systems continues. The following section provides an industry perspective on the net-centric transformation that captures a number of the sentiments expressed within this chapter.

## INDUSTRY PERSPECTIVE ON THE NET-CENTRIC TRANSFORMATION

*This provides a perspective from an industry veteran as to how the net-centric transformation is affecting the corporate world and creating new markets. Industry efforts are continuing on subjects such as*

*"Net-Centric Computing" which takes advantage of military research and uses it for the corporate IT field. By Russell Boyd*

Mr. Russell S. Boyd serves as the Practice Lead for Enterprise Architecture and Investment in the Defense marketplace with an emphasis on enabling government agencies to achieve IT legacy modernization, business transformation and information sharing objectives. Mr. Boyd is a retired Air Force officer who was active in the Air Force Medical Corps.

A variety of emergency situations in recent years have demonstrated in increasingly vivid detail the tragic consequences that often result from the inability of jurisdictions and agencies to effectively share information. Terrorist attacks, natural disasters, and large-scale organized criminal incidents too often serve as case studies that reveal weaknesses in our nation's information sharing capabilities. Moreover, enterprise-wide information sharing is also required to support the critical day-to-day operations of public safety officials at all levels and across all branches of government.

Most organizations have a large collection of information systems to support many disparate functions. The majority of these applications were originally built as individual, stove-piped solutions to support very specific low-level processes. As computer-intensive applications have become more pervasive, the ability of organizations to meet their strategic objectives hinges on their ability to quickly, accurately, and effectively access, analyze and process data. However, legacy systems often make reaching these objectives more challenging because they are hampered by inconsistencies and inefficiencies resulting from integration difficulties due to incompatibilities across application control and data ownership.

The Internet has facilitated an e-commerce explosion in the private sector through the ready access of information and business services; it has also helped companies lower costs across supply and demand chains, dramatically improve customer service, and redefine business relationships. Current information collection and dissemination practices have not been planned as part of a unified national strategy but rather have evolved incrementally over time to meet certain needs or address specific challenges as they have surfaced. While sharing does occur through these stove-piped efforts, it is commensurately limited in degree and effectiveness. A tremendous quantity of information that could be shared is still not effectively shared and utilized among communities of interest (COIs).

There is no simple solution that will fix all the challenges the U.S. faces with respect to information sharing, but net-centric computing and information sharing are addressing the issue.

## The Role of Net-Centric Computing

The underlying principle of Net-Centric Computing (NCC) is a distributed environment where applications and data are downloaded from servers and exchanged with peers across a network on as as-needed basis. This is in stark contrast to the use of powerful personal computers that rely primarily on local resources. In some respects, NCC resembles an earlier computing era of mainframes and dumb terminals. However, there are significant differences. NCC relies on portable applications running on multiple platforms, mobile data accessed via high-speed network connections, and low-cost appliances for local processing.

The last few years have seen an interest in net-centric computing increase, decline, and then grow again. At the moment, there appears to be a resurgence of interest in the area, driven in part by the proliferation of non-traditional computing devices, by dramatic changes in networking capabilities, and by renewed interest in centralized system administration. Corporations are currently using NCC technologies to assist with the problem of enterprise integration. The potential of the Web as a portal to corporate data provides a mechanism to allow for searching and retrieval of inquiries, as well as for access to information about orders, order fulfillment and customer data.

Early uses of the Web using HTML permitted static information to be displayed, but did not provide benefit for enterprise integration. Later, the introduction of Java provided for portability across hardware platforms and operating systems. Technologies such as Enterprise JavaBeans (EJB) simplified the development of middleware components that support transactions, security and database connectivity.

More recently, emerging internet standards such as XML provide the capability to integrate the structured data from databases and legacy systems with the unstructured data of text documents, graphics and images. These capabilities can facilitate the integration of legacy systems, as well as provide for the integration required for emerging e-business and B2B applications. While the use of Web technologies to support information sharing has strong potential, Web technologies increasingly leverages shared services and Service Oriented Architecture (SOA) applications.

Achieving the commitments in Information Sharing and NCC requires a fundamental change in the way Information Technology (IT) is provided and managed by corporations today. Historically, IT resources and software-based capabilities have been acquired and managed as stand-alone systems rather than as integral parts of a net-centric capability. System-to-system connections are defined, engineered, and implemented one pair at a time. This approach focused on system or platform capabilities rather than on mission or business capabilities. The result is multiple overlapping implementations, limited ability to share information, and a rigid set of capabilities that are unresponsive to the commercial world's evolving needs.

The commercial world's approach to resolving this problem has been to define business processes as workflows. These workflows consist of specific business functions that are supported by the delivery of software-based services over networks. These software-based services deliver reusable business functionality as standardized building blocks on an enterprise network.

Implementing NCC interoperable and integrated technology to support corporation mission processes is critical. In order to accomplish complete NCC corporations need to include the effort to improve information sharing across the commercial business enterprise.

## Information Sharing Overview

Learning from military net-centric projects, there are three core tenets of information sharing: strategies and drivers, architecture, and governance. Strategy and drivers influence the direction a corporation is taking with information sharing and the initiatives it is involved in. Corporations should consider three objectives to improving information sharing, including: share information across the extended business mission Enterprise, develop and implement required information sharing, data security and privacy policies, and develop information sharing architecture standards.

Enterprise Architecture is needed because it describes the "as-is" and "to-be" state of the corporation's performance, business, data, services and technology architectures. Corporations need to align their enterprise architecture to other industry frameworks and reference models such as the Federal Enterprise Architecture (FEA) and DoDAF to provide a common vocabulary for sharing information, a standard structure for developing exchanges, and enable publication and facilitates search and analysis of information. Once the EA is established, corporations need to invest in developing segment architectures. Segment architecture is a focused subset of the EA that is consistent and compliant with the FEA and concentrates on getting results for a specific mission area, common business service or enterprise service.

A comprehensive, widely supported governance framework is critical for creating and sustaining a federated information sharing community of like corporations, agencies and individuals aligned to the information sharing vision and agreed upon rule sets and processes. Successful information sharing also requires sustained leadership, commitment, and determination; renewed and strengthened cooperation between agencies; constant evaluation and assessment; and, when necessary, the introduction and adoption of new ideas and approaches. Communication, outreach and training are crucial to gaining the buy-in and active participation that is needed to successfully foster information sharing.

## REVIEW OF CHAPTER GOALS

The goals of this chapter were to address:

1. How are the command and control systems set up that address decision making in a hierarchical manner:
   - Provide typical characteristics of command and control systems
2. How has the civilian and government contracting industry shifted toward Net Centricity:
   - Address the industry roadmap that includes a semantic infrastructure, sensory-based systems, knowledge management and integrated cognitive systems
3. What types of DoD domestic transfer policies have been set up for military to civilian technology transfer:
   - Discuss Cooperative Research and Development Agreements (CRADAs)
4. How can we pictorially depict the elements of the defense information enterprise:
   - Discuss each of the organizational element of the overall enterprise
5. What types of industry topics are related to Net Centricity:
   - Describe networking and communications, web technologies, network security, multimedia technologies, and mobile computing
6. What are the upcoming industry technology areas within this space:
   - These would be building a semantic infrastructure, cognitive systems and the growth of net-centric computing

## REFERENCES

Alberts, D. S., Garstka, J. J., & Stein, F. P. (1999). *Network Centric Warfare: Developing and Leveraging Information Superiority,* (2nd Ed.). Washington, DC: CCRP Publication Series.

Alberts, D. S., & Hayes, R. E. (2005, April). *Power to the Edge.* Command and Control Research Program Publication Series. Available at www.dodccrp.org

Bridges, S., Zeigler, B. P., et al. (2005) *Evolving Enterprise Infrastructure for Model & Simulation-Based Testing of Net-Centric Systems.* White Paper, Joint Interoperability Test Command, Ft. Huachuca, AZ.

Department of Defense Joint Publication 1-02 (2001). *DoD Dictionary of Military and Associated Terms*. Washington, DC: Author. Retrieved from http://www.dtic.mil/doctrine/jel/new_pubs/jp1_02.pdf

Eleazer, B., & Reeder, S., (2007, April). *Testing Concept of Operations (CONOPS) in DoD's Net Centric Environment Testing.* 8thAnnual NDIA Science & Engineering / DoD Tech Expo.

Newman, A. V., & Wheelwright, R. W. (2006) *A Proposal For A Department Of Defense (DoD) Command And Control Strategy*, International Command and Control Research and Technology Symposium

Office of the Assistant Secretary of Defense for Networks and Information Integration, Department of Defense Chief Information Officer Directive 8320.02-P. (2007, April). *Data Sharing in a Net-Centric Department of Defense.*

Office of the Assistant Secretary of Defense for Networks and Information Integration, Department of Defense Chief Information Officer. (2008). *Department of Defense Information Management/ Information Technology Strategic Plan 2008-2009.*

# Chapter 7
# Healthcare Transformation in a Net-Centric Environment

**Mark Janczewski**
*Consultant, USA*

## CHAPTER CONTENT

As you explore through Chapter 7, it covers the following topics:

- Overview of Military Medicine and Veteran Care
- The Continuum of Patient Care
- Net-Centric Transformation of Military Medicine
- CASE STUDY: Theater Medical Information Program
- Transformation Perspective on Net-Centric Medicine

## CHAPTER FOCUS

As a transformative perspective, this chapter discusses a specific area of DoD that is undergoing a large transformation to provide healthcare to today's military and engage with the Department of Veterans Affairs to take care of veterans. We provide an overview of military medicine and veteran care, and dis-

DOI: 10.4018/978-1-60566-854-3.ch007

cuss the important concept of the patient care continuum. We then discuss a continuum of care scenario for wounded warriors from DoD to VA. We describe the type of transformation that is being conducted in the Military Health System, and then provide a case study of a military health program known as Theater Medical Information Program (TMIP). We end with a transformation perspective on net-centric medicine provided by a retired DoD medical officer.

In this chapter, as part of our section objective to address net-centric transformation perspectives, we provide a synopsis that specifically caters to the healthcare for our military warfighters and veterans. This then leads to the discussion of DoD Military Health System which is currently trying to transform itself to address Net Centricity. As part of this discussion, we describe the DoD Continuum of Care that happens for a warfighter from the point of injury in the battlefield all the way to him receiving veteran care. We use the Theater Medical Information Program (TMIP) as a net-centric example case study to address military healthcare requirements and the Net-Centric Key Performance Parameter mandates. The chapter then provides a guest author perspective on the DoD and VA healthcare transformation.

## OVERVIEW OF MILITARY MEDICINE AND VETERAN CARE

The Department of Defense and Veterans Affairs (VA) federal agencies have been working together for many years to ensure that our nation's soldiers and veterans are provided with government funded medical and rehabilitative care. Both agencies have been working based on the following cooperative principles:

- Establish goals for the Continuum of Care that are patient centered and effective in delivering world class care
- Provide care that is efficient without unwarranted burden on government resources
- Provide care that is needed and deserved at the right time, right place and is provided by the right person
- Promote the successful rehabilitation and reintegration of service members; organize and leverage all inter service and interagency resources
- Provide service members with the responsibility to make choices about their care.

Based on these cross-agency principles, our government has been trying to provide the most effective care to its soldiers, veterans and their immediate beneficiaries. Since medical care for our soldiers span their entire career including their time as a veteran, this span is known as the Continuum of Care.

In this chapter, we try to address the ongoing concerns of military health and its impacts, as the DoD transforms toward a net-centric future. It is best then to provide the overall goals and mission of the Military Health System (MHS) and Veterans Health Administration (VHA) agencies, so that we can come to a better understanding of the issues.

### Goals of Military Health System

The DoD Military Health System (MHS) has a two-part mission. These two parts can be stated as separate goals.

- **Goal 1:** To provide and maintain readiness, to provide medical services and support to the armed forces during military operations.
- **Goal 2:** To provide and maintain health and medical services to support members of the armed forces, their dependents, and others entitled to Department of Defense medical care.

Within the DoD, healthcare is a joint activity that then is coordinated with each of the medical commands of the Secretary of Army, Navy and Air Force. MHS efforts are under the TRICARE Management Activity (TMA) and are based on the direction of the Assistant Secretary of Defense (Health Affairs) (ASD (HA)) and serves as one of the federated components within the DoD enterprise. The first goal is to address the healthcare of the soldier within the Theater and battlefield, and then with the military commands. The second goal is to address the healthcare of the soldier and their dependents and beneficiaries within the sustaining base in fixed hospitals throughout the United States.

Based on the MHS mission, the overall government and military objectives are to:

- Respond anytime, anywhere with comprehensive medical capability to military operations, natural disasters and humanitarian crises around the globe
- Ensure delivery of healthcare to all DoD service members, retirees, and their families
- Promote a fit, healthy and protected force by reducing non-combat losses, optimizing healthy behavior and physical performance, and providing casualty care

## Goals of Veterans Health Administration

Veterans Health Administration (VHA) is one of three primary components of the U.S. Department of Veterans Affairs. VHA is a federally funded and centrally administered health care system for veterans. The two other primary components of the VA are to administer on behalf of veterans, a program for disability benefits and a national system of cemeteries.

VHA has a four-part mission: patient care, research, teaching, and contingency backup for the DoD medical care system.

- **Patient Care** – VHA is one of the country's largest healthcare delivery systems, and based on 1999 data, the entire system included 172 hospitals, 132 nursing homes, 73 home health care programs, 40 residential care programs, and more than 600 outpatient clinics
- **Research** – The VHA has an extensive research and teaching activity. VHA manages research programs in biomedical sciences, rehabilitative medicine, and health services delivery systems.
- **Teaching** – VHA fulfills its teaching mission based on academic affiliations with the country's medical schools and schools of allied health professions. VHA is integrated with the country's system for graduate medical education, and provides financial support and clinical training to approximately one-third of the country's medical residents.
- **Contingency backup for the DoD** – As part of VHA's role, it has two primary responsibilities. One is to support the DoD military healthcare during times of war. The other is to assist the Public Health Service and the National Disaster Medical System in providing emergency care to victims of natural and other disasters.

## THE CONTINUUM OF PATIENT CARE

In this section, we provide a glimpse of the continuum of care as it stems from the battlefield point of injury all the way to rehabilitative care. As citizens, we all want to make sure that our sailors, airmen and soldiers return to duty, to their lives, and to their families after a traumatic injury in the battlefield. This is a very large task. It is a combined effort to make sure that the wounded, ill and injured is taken care of by the DoD, the VA, and a network of civilian partnerships such as hospitals and clinics. It fast becomes clear that this patient care continuum is largely dependent upon the efficiency of the supporting information management systems that is currently deployed by the MHS and VHA.

### Continuum of Care Scenario

For the wounded soldier who makes the journey from battlefield through the healthcare system, every stop he or she makes generates information that is critical to the success of the continuum of care, and the care that they receive. We have created an illustration that can be labeled a DoD/VA Continuum of Care scenario for a wounded warrior returning from the battlefield. This provides a notional path as to how information flows from the point of injury in the battlefield across the DoD Levels of Care and transition to VA healthcare services.

#### DoD Levels of Care

As we discuss the plight of the wounded warrior, it is important to discuss the existing levels of care that is offered by the DoD for addressing the medical needs of the soldier. The following levels of healthcare are provided within the DoD for its military servicemen.

- **Level I Care** – Level I care consists of care rendered at the unit level at the point of injury. It includes self-aid, buddy aid, and combat lifesaver skills, examination, and emergency lifesaving measures such as the maintenance of the airway, control of bleeding, prevention and control of shock, splinting or immobilizing fractures, and the prevention of further injury.
- **Level II Care** – At a minimum, Level II care includes physician-directed resuscitation and stabilization and may include advanced trauma management, emergency medical procedures, and forward resuscitative surgery. Supporting capabilities include basic laboratory, limited x-ray, pharmacy, and temporary holding facilities. Patients are treated and returned to duty, or are stabilized for movement to a MTF capable of providing a higher level of care.
- **Level III Care** – Level III care is administered that requires clinical capabilities normally found in a facility that is typically located in a reduced-level enemy threat environment. The facility is staffed and equipped to provide resuscitation, initial wound surgery, and postoperative treatment. This level of care may be the first step toward restoration of functional health, as compared to procedures that stabilize a condition to prolong life.
- **Level IV Care** – In addition to providing surgical capabilities found at Level III, this level also provides rehabilitative and recovery therapy for those who can return to duty within the theater patient movement policy. This level of care may only be available in mature theaters.
- **Level V Care** – Level V definitive care includes the full range of acute convalescent, restorative, and rehabilitative care and is normally provided in CONUS by military and Department of

*Figure 1. DoD/VA Continuum of Care Scenario for a Wounded Warrior*

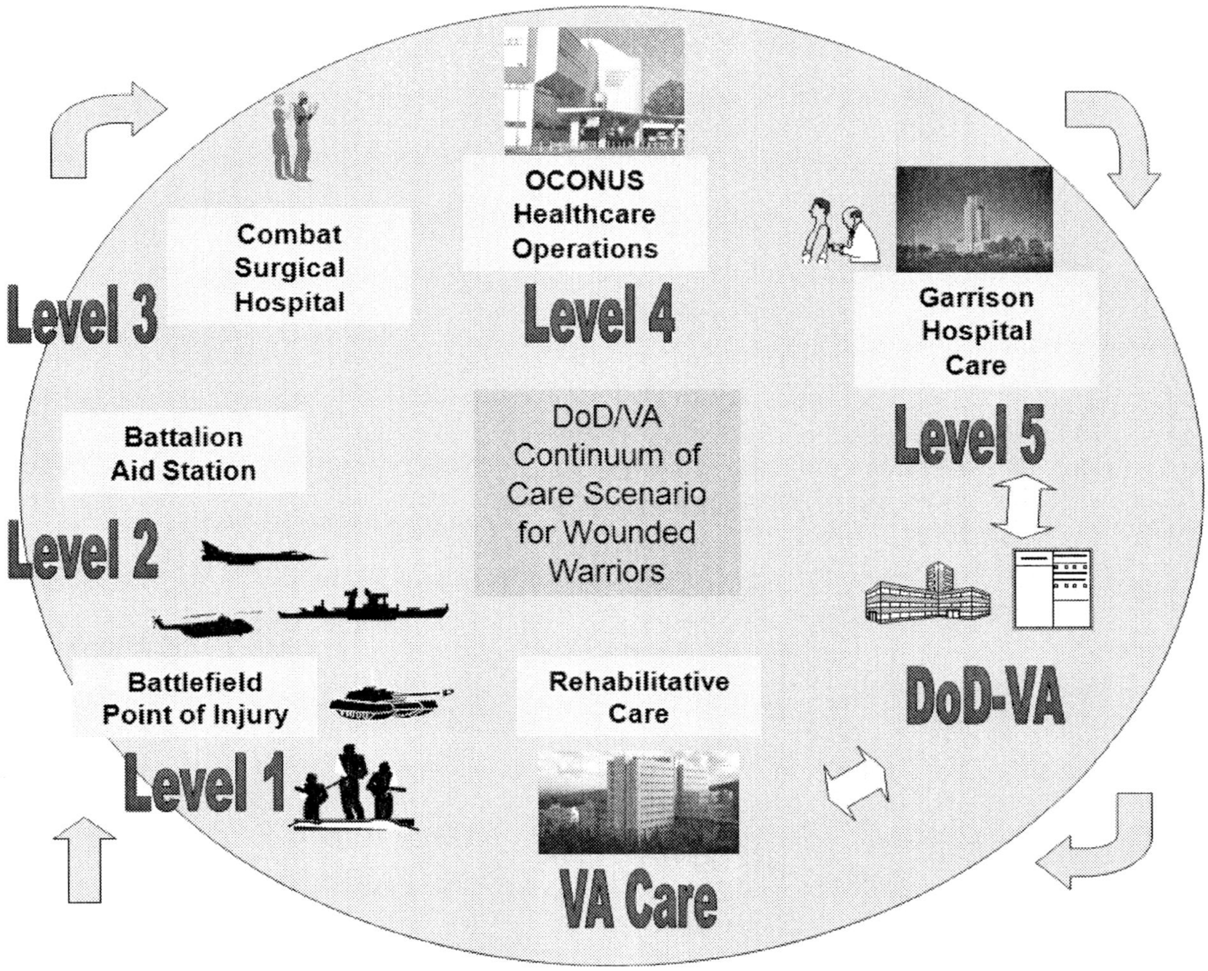


Veterans Affairs hospitals, or civilian hospitals that have committed beds for casualty treatment as part of the National Defense Medical System. The DoD Levels of Care is then continued by VHA based on collaboration between the DoD and VA to appropriate the necessary support and rehabilitative care provided by VA medical centers and the VA healthcare system.

This is shown in a pictorial manner in Figure 1 as a continual depiction of care.

## Example of Wounded Warrior Care

In illustrating the continuum of care for our wounded warriors, we offer one example, laid out in a step-wise manner, of how healthcare takes place for one soldier who makes a journey from the battlefield through the military healthcare system, the VA healthcare system, then final ongoing treatment and rehabilitation near his home.

In illustrating the continuum of care for our wounded warriors, we offer one example, laid out in a stepwise manner,

- A solider is severely injured by a roadside bomb in Iraq. Based on Level 1 care, he receives first aid in the front line.
- He is then provided surgical resuscitation at the battalion aid station as Level 2 care.
- He is then immediately transported to Balad AFB in Iraq and to the 10th Combat Support Hospital, or Level 3 care, where his immediate wounds are addressed.
- But since he has traumatic brain injury from the blast, he requires immediate evacuation and aggressive specialized treatment, if he is survive. Within 48 hours, he is picked up by an Air Force transport team and transferred to Landstuhl Regional Medical Center (LRMC) in Germany. In this Outside of Continental U.S. Level 4 facility, his medical care providers review his past medical history, and update his records within the DoD Electronic Health Record.
- He is then prepared for his journey back home to Continental United States military treatment facility. He arrives at Walter Reed Medical Center, where he has definitive care, surgery and rehabilitation waiting for him.
- Since he is not able to be on active duty because of his injuries, he is then separated from the military and given Combat Veteran status, and he receives further care at a VA medical center near his home.
- Other private hospitals and clinics near his home also become involved, since he also goes to them and he coordinates his healthcare benefits with his dependents and beneficiaries.

The journey that this wounded warrior takes on his way to recovery represents a network of business processes, information technology systems, geopolitical influences, and funding sources that all have to work together to restore the health of the soldier and his beneficiaries. The MHS transformation for continuity of care works on the concept of continuity of information. This refers to the MHS ability to provide information that enables a seamless transition and escalation of care across multiple agencies, Service medical departments and third-party networked healthcare providers.

As we investigate the information systems that are involved in providing healthcare, we recognize that today many of these information exchanges are paper-based. As part of the MHS transformational efforts of tomorrow, the MHS is looking forward to a joint electronic health record that provides interoperability to healthcare. This is in line with the larger DoD net-centric effort to prepare the military for the future.

The soldier's entire longitudinal health record is captured and stored electronically so that it provides a full medical history to any and all medical specialists. The upcoming transformation would fully implement a joint electronic health record that allows DoD to transfer the wounded warrior's electronic record directly to the VA and to third-party providers that are all part of civilian hospital chain.

DoD is also participating in the National Health Information Network (NHIN), which is an initiative to allow the federal government to develop national standards for the exchange of health information. The coordination between government agencies such as DoD, VA, and Health & Human Services (HHS) with regard to medical electronic exchange. It is designed to fill in the gaps between all of the military health partners to ensure continuity of care through the continuity of information.

## DoD MTF Care

DoD's Military Treatment Facilities (MTFs) is the focus of its health care delivery in the sustaining base and for peacetime. DoD operations in OCONUS and CONUS are part of the span of MTFs that provide care across the globe. The medical mission of DoD is to provide and maintain readiness, medical services, and support to the armed forces during military operations and to provide medical services and support to members of the armed forces, their family members, retirees and their families, and eligible survivors of deceased active and retired military personnel.

For the wounded warrior, DoD's health care program provides a range of medical services such as surgery and inpatient care, pharmacy services, and mental health care to eligible beneficiaries. This care is delivered through the military hospitals and clinics nearby to a soldier, or from contracted civilian-provided care.

### Transition to VA Healthcare Services

VA's goal is to ensure that every seriously injured or ill serviceman and woman returning from combat receives easy access to benefits and healthcare services. Combat veterans have special health care eligibility. Their contact with the VA often begins with priority scheduling for care. For the most seriously wounded, VA counselors visit the bedside of wounded warriors in military wards before separation to ensure that their VA disability payment coverage will be ready the moment they leave active duty. VA employees based at military treatment facilities brief service members about their VA health benefits, disability compensation, vocational rehabilitation and employment. At the VA facilities serving the veteran's home town, the hospital is alerted when the seriously wounded service member is being discharged so that the continuity of his or her medications and therapy is ensured when they arrive home.

## NET-CENTRIC TRANSFORMATION OF MILITARY MEDICINE

Based on the continuum of care scenario, it makes sense that a transformation needs to take place if each of the steps within this healthcare process has to be done electronically and in an automated manner. This change in the healthcare process will also affect the military posture with regard to readiness of the military to perform mission critical tasks. Hence, the healthcare transformation is a subset of the larger net-centric process that the overall DoD is undergoing.

The Military Health System (MHS) transformation vision is to develop a collaborative, agile, and efficient medical business enterprise that adapts to the changing needs of military medicine and maximizes the benefit of business and IT resources. The MHS business objective focuses on continuity of care across the DoD and VA and the civilian healthcare delivery system. This intends to be a shift from reactive to proactive care which then will hopefully result in more efficient healthcare operations.

The current MHS business transformation strategy is driven by three pillars of the overall MHS strategic plan. The expectation from these pillars is the following:

- MHS will provide medically-ready force protection for military communities through the implementation of medical surveillance and health monitoring solutions and processes. The intention is to improve the health of the soldier, enhance human performance and protect the community against medical threats across the entire range of military operations.
- MHS will create a deployable medical capability to address theater operations, which is enabled by globally accessible healthcare information. This envisions a rapidly deployed set of medical services and products that can go anywhere, and anytime with flexibility, interoperability and agility.

- MHS will manage and deliver a health benefit by strengthening the DoD partnerships with the military beneficiaries, and coordinating care with other federal health agencies including the VA, HHS, and private sector hospitals and clinics.

The MHS works to provide healthcare services and support structure that serves the DoD's 9.1 million active duty and reserve personnel, family members, and retired beneficiaries. TRICARE Management Activity also supports each of the Military Services that includes the Army, Navy, and Air Force Surgeon Generals, to maintain medical readiness and to provide medical and dental services to members of the Armed Forces during military operations.

## Military Health Business Transformation

MHS is transforming business practices to optimize the integration, efficiency, and effectiveness of the DoD healthcare system. This transformation is in line with our central goal within this book, which is to promote the long-term vision of Net Centricity. The MHS will realize this transformation through the implementation of the following goals and objectives:

- **Provide continuity of care through continuity of information**
  - MHS plans to implement processes and information solutions that helps to ensure that no matter:
    - where a patient is located
    - or what healthcare provider is treating them
    - that information and medical material products and services are readily available at the point of care
  - In addition to deploying greater functionality in the DoD Electronic Health Record (EHR) and medical logistics suite of applications, MHS needs to integrate critical components of the federal Nationwide Health Information Network (NHIN)
  - This care spans DoD MTFs, VA and other federal health agencies, and a large number of private sector "network" providers that serves DoD's medical beneficiary population
- **Transform from a reactive to a proactive healthcare system**
  - Keep patients healthy and active within the community to take a proactive vs. reactive approach to healthcare
    - Keeping patients healthy is more than just caring for them once they become sick or injured
  - Understand the factors that threaten health in the first place, and anticipate the needs of our patient and provider teams.
    - By addressing potential health issues before they become real problems, the intention is to have patients and providers partner to prevent disease rather than react to it.
  - The collection, analysis, sharing, and application of health data not only enables proactive healthcare on a patient by patient basis, but also can make possible access to a wealth of medical information that can assist decision-makers from a global perspective.
    - As patients flow through the medical system along the continuum of care, they generate information every time they encounter their healthcare teams.

        - Patient encounter information accumulated in EHRs allows DoD to continuously monitor individual health status; recognize and track trends, outbreaks, and exposures to hazards
    - This contributes to finding new ways to protect and improve the health of individuals, communities, and our nation
- **Enhance military health benefit through more efficient healthcare operations.**
    - The promotion of medical efficiency is a hallmark of quality
    - Quality medical coding contributes to the efficient processing of claims and contributes to the efficiency of our medical surveillance
    - Quality patient safety controls help to get patients out of the medical system quickly, and protect them from medical errors and mistakes
    - Quality medical logistics enables smooth flow of materiel and pharmaceuticals to, from and between patients, providers, suppliers and intermediaries

## CASE STUDY: THEATER MEDICAL INFORMATION PROGRAM

We now provide a case study of an actual DoD program that is currently under development that has to abide by net-centric mandates set forth by DoD directives and instructions. This program is a crucial element within the MHS and serves to tie the medical theater of operations to fixed facility military hospitals operating during peacetime.

### TMIP-J System of Systems

The Joint Theater Medical Information Program (TMIP-J) is a DoD program that is the medical component of large joint systems known as Global Command and Control System (GCCS) and Global Combat Support System (GCSS). This joint DoD system of systems provides an Army, Navy, and Air Force component of the TMIP program. The TMIP-J system of systems follows a federated architecture that provides the following functionality:

- It is a set of medical information systems used in the sustaining base to support peacetime requirements that are tailored and integrated to provide automated information management support in a theater environment
- The program deploys on Armed Force Service-provided computer hardware and communications infrastructure in accordance with individual Service requirements
- TMIP-J is intended to be interoperable to other MHS information systems such as the overall DoD Electronic Health Record (EHR)
- The program intends to provide the capability as authorized by the Combatant Commander to transfer all medical data gathered during military operations to appropriate MHS records and systems

### TMIP-J Net-Centric Capability

TMIP-J is required to provide a net-centric capability in order for the DoD to abide by the following critical military needs:

- Ensure medical situational awareness information required for medical surveillance for the readiness of troops during pre-deployment, in the theater and post-deployment
- Provide medical infrastructure management within the theater that is then available to decision-makers when and where they need it

## TMIP-J Operational Capability

TMIP-J operations support the DoD Force Health Protection (FHP) activities. FHP encompasses all of the measures taken by commanders, leaders, individual Service members, and the MHS to promote, improve, conserve, or restore the mental and physical well being of Service members across the range of military activities and operations. TMIP-J operations support the following program functions:

- Maintenance of a healthy and fit force
- Casualty prevention and casualty care management in operational theater
- Medical surveillance and intelligence data needed for protection from all health threats across the full range of military operations
- Tailored, standardized joint medical systems to provide only essential care in theater
- Enhanced care during evacuation to definitive care
- Improved health monitoring and surveillance of forces engaged in military operations
- Improved medical situational awareness across interfaces
- Improved patient tracking that includes the ability to access and transmit medical data in real time

## TMIP-J Data Strategy

TMIP-J data strategy is to automate the collection of patient encounter data at the point of care. This means that when the injured soldier or servicemen show up to the medical team, the system intends to capture and store computerized data about occupational and environmental health hazards and patients' health status and healthcare, and make this information accessible to authorized users across the DoD. The system data provides medical logistics functionality and the ability to monitor and analyze health trends among the population-at-risk and to formulate risk management strategies for appropriate interventions. The system has to be able to request tactical and strategic patient movement and the ability to track patients while they are in-transit between medical treatment facilities.

This broad data strategy needs to be configured for the net-centric future by abiding by net-centric core principles. TMIP-J data supports medical command and control users at the Service intermediate headquarters levels, the Service component, the Joint Task Force Commander and the Combatant Commander levels. The data parameters and data goals for the system are the following:

- **Data Accuracy:** Data stored in the TMIP-J database must accurately reflect the information accepted and must maintain that accuracy over time.
- **Data Completeness:** TMIP-J shall maintain and guarantee during transport the integrity of all information elements exchanged throughout the GIG to enable user confidence.
- **Data Accountability:** TMIP-J has the ability to amend and append clinical documentation. Provide an audit trail showing the source and time of any system changes to data.

*Table 1. MIP-J Key Performance Parameters*

| | |
|---|---|
| **Medical Infrastructure Readiness** | TMIP-J shall provide the capability to document, access, aggregate and display information to support assessment of medical personnel, medical units, medical facilities, medical equipment and supplies to determine capabilities and limitations during deployment and sustainment. System shall have the capability to acquire needed information by search, with successful searches yielding 95% accurate data from the authoritative source; data is current as determined by established business processes. |
| **Patient Visibility** | TMIP-J shall provide current, accurate and reliable patient visibility information during contingencies from initial point of care within and between theater MTFs, and during movement within the theater and between theaters enroute to CONUS MTFs. Patient visibility applies to U.S. Service members and contractors, allied and coalition forces, enemy prisoners of war, and host nation and local national civilians. System shall have the capability to acquire needed information by search, with successful searches yielding 95% accurate data from the authoritative source; data is current as determined by established business processes. |
| **Medical Surveillance** | TMIP-J shall provide access to medical threat and provide the capability to access, aggregate and display medical surveillance information.<br>TMIP-J shall provide the capability to access, aggregate and display medical readiness status for individuals and for units in the theater in accordance with Assistant Secretary of Defense (Health Affairs). System shall have the capability to acquire needed information by search, with successful searches yielding 95% accurate data from the authoritative source; data is current as determined by established business processes. |
| | TMIP-J shall provide ability to access, aggregate and display illness and injury rates and trends for the theater in accordance with Office of the Chairman Joint Chiefs of Staff. System shall have the capability to acquire needed information by search, with successful searches yielding 95% accurate data from the authoritative source; data is current as determined by established business processes. |
| | TMIP-J shall provide the capability to access, aggregate and display rates and trends for Tri-Service reportable events data in accordance with Office of the Chairman of the Joint Chiefs of Staff. System shall have the capability to acquire needed information by search, with successful searches yielding 95% accurate data from the authoritative source; data is current as determined by established business processes. |
| **Medical Logistics Visibility** | TMIP-J shall provide the capability for authorized users to access current and accurate location information for medical (class VIII) supplies and equipment, and for blood and blood products. Includes ITV within and between theaters, and the location and quantities of on-hand stocks within the theater. System shall have the capability to acquire needed information by search, with successful searches yielding 95% accurate data from the authoritative source; data is current as determined by established business processes. |
| **Data Transfer** | TMIP-J shall provide the capability to transfer all medical data gathered in theater to appropriate records and record keeping systems in the sustaining base with 99% accuracy when authorized by the Theater Combatant Commander. Includes documentation for: inpatient and outpatient medical and dental care; immunizations administered in Theater; pre- and post-deployment surveys; industrial hygiene documentation; and hearing conservation documentation. |
| **Standards Compliance** | TMIP–J shall be compliant with the DoD Information Technology Standards and Profile Registry (DISR). |
| | TMIP-J shall be compliant with applicable GIG requirements. |
| | TMIP-J shall provide information assurance and security management services. |
| | TMIP-J shall protect against unauthorized disclosure of both personal privacy and patient health data. The system shall handle data in accordance with Public Law 104-191 Health Insurance Portability and Accountability Act (HIPAA) of 1996 – Protecting Electronic Health Information. Data to be protected shall be registered in accordance with DoD Discovery Metadata Registry policy in the DoD Metadata Repository. |
| **System Operational Availability** | System operational availability shall be 99%. |
| **Net-Ready KPP** | All interfaces; services; policy-enforcement controls; and data correctness, availability and processing requirements designated as enterprise-level or critical in the Joint integrated architecture.<br>All activity interfaces, services, policy-enforcement controls, and data-sharing of the NCOW-RM and GIG-KIPs will be satisfied to the requirements of the specific Joint integrated architecture products (including data correctness, data availability and data processing), and information assurance accreditation, specified in the threshold (T) and objective (O) values. |

- **Data Recovery and Restoration:** Recovery and restoration must be achievable following a TMIP-J hardware, application, and database or power failure.
- **Data Integrity:** The integrity of the data (i.e., accuracy, currency, and format) must be maintained when the data is retrieved and used by TMIP-J applications and when it is accessed in a shared data environment.
- **Input Edit Checks:** TMIP-J input edit checks must identify to any user attempting to enter data, any non-conformance with pre-established data field boundaries and reject non-conforming entry attempts.
- **One-time Data Entry:** Once discrete patient and clinical data passes input edit checks and is entered into the TMIP-J database, no further entry of the same data into the database will be required.

### TMIP-J Key Performance Parameters

The following table has been excerpted from TMIP-J program requirement specifications as was written in the TMIP Capability Production Document (TMIP CPD Increment 2, 2004) as milestone documentation. This provides you with a real understanding of the key performance parameters that are necessary to make the program successful. This includes specific implementation of net-centric parameters for the DoD medical domain. These parameters have threshold and objective values assigned by senior leadership that the program has to meet during its lifetime.

## TRANSFORMATION PERSPECTIVE ON NET-CENTRIC MEDICINE

*Perspective from a DoD medical officer that provides an understanding of current DoD healthcare systems and an upcoming glimpse into Net-Centric Medicine and sharing of information between Military Health System and Veterans' Health Affairs along with other federal agencies. By Dr. Mark Janczewski, Air Force Colonel (ret.), MD, MPH*

Dr. Mark G. Janczewski is an expert in military medicine and health information technology. He is a board-certified physician and has an extensive medical and educational background having received numerous academic degrees. He is a retired Colonel from the U.S. Air Force and currently consults for the DoD and Department of Veterans Affairs.

The Military Health System (MHS) is an immense healthcare enterprise supporting over 9 million beneficiaries ranging from active duty military personnel, to dependents (spouses, children, etc.), to retirees. There is a significant challenge of providing these services worldwide, since the MHS is arguably the most geographically spread-out healthcare system in the world. Further, most of its active duty provider force and a large percentage of beneficiaries relocate every 2-3 years. It is best to note that the overall mission of the MHS is to first ensure a medically fit military force, especially those servicemen and women serving in austere or hostile environments and also to provide healthcare to other beneficiaries within the MHS's purview.

The intent of the MHS is to provide a longitudinal Electronic Healthcare Record (EHR). The word "longitudinal" denotes a lifespan of primary medical care, laboratory results and pharmacy drug interactions. The challenge is that to ensure the highest quality of medical care, military health care providers need to have 24/7 access to the longitudinal clinical record of each patient seen. In the past, this daunting challenge was met through paper records, and whenever a military family moved, all family member medical records were shipped to the gaining military base medical facility; this resulted in a process that was not always 100% reliable.

By the 1990s, most military treatment facilities (MTFs) had access to a computerized physician order entry (CPOE) software program known as Composite Health Care System or CHCS. It allowed providers to obtain laboratory results and radiology reports, and order medications. For administrators, patients could be scheduled using this system. The system was based on the MUMPS language. While considered very useful for healthcare personnel, it did not have a Windows-like graphical user interface; it was more like working in MS-DOS character-based environment. Results retrieval and medication ordering could only be accomplished within the particular MTF that the soldier was assigned, or in some cases, a local group of MTFs utilizing one server. For example, the National Capital Area (NCA), which included Army, Air Force, Navy and Marine MTFs in the Washington, DC metropolitan area, did not have access to a common server and database until after 2000. This presented a problem for many patients who may have received their primary care in one facility, such as Andrews AFB, and then specialty care at Walter Reed or Bethesda Naval hospitals. Thus, it was not possible to see all of the medications a patient was currently taking by reviewing prescriptions obtained from clinicians at the different facilities.

Conceptualization, design and development of the next "generation" of information systems to support the healthcare provider began in earnest in the 1990s with CHCS-II. Its aim was to provide a longitudinal electronic health record that would be available to all healthcare providers in all MTFs whenever needed. This was a significant step forward from two respects: first, the application now had a GUI which was fairly easy to navigate, and second, there was a central clinical data repository (CDR) that stored patient health data obtained from a large number of host MTFs. To help "populate" this CDR from the outset, whenever CHCS-II was installed at a new MTF, a 25-month "data pull" was accomplished to retrieve data from the local CHCS (legacy) host. By 2004, CHCS–II was renamed to different moniker, called AHLTA, and this composite system was deployed to all MTFs.

It has taken time for AHLTA to take hold and garner praise from the military healthcare provider community; yet one of its major contributions has been realized – providing a robust, reliable and readily available longitudinal health record. While some may have concern about its complexity of use, AHLTA provides a very real example of the concept of Net Centricity in the healthcare setting. It has proven to provide an extremely rich repository of data whose value for other areas such as medical outcome studies is quite promising. As the knowledge repository grows, it is providing a new set of untapped possibilities.

Today, the AHLTA system in conjunction with the use of AHLTA-T (for "theater"), is providing an invaluable mechanism for the continuity of clinical information flow for our military personnel in deployed environments from Iraq, Afghanistan and neighboring countries. Warfare unfortunately exacts a human toll in lives lost or wounded. Having a tool available to readily move patient data from one location to another, as the patient moves, is one aspect of AHLTA's net-centric capabilities that needs to be tested and understood.

Other information systems that can ride on top of the main electronic health record such as those that allow global awareness and access of digital medical images and for inventorying and ordering facility supplies are either in development or being further refined. We hope that eventually, these systems will also become interoperable to the extent necessary.

Some particular examples of the usefulness of this net-centric architecture include the following:

- MTF manning levels for clinical and administrative personnel is based on workload; workload is based on the number and extent of patients seen. Extracting workload reports based on AHTLA patient encounter data can help provide this information to decision makers, thus enhancing the flexibility of military medical commanders who often quickly need to reallocate human resources.
- Knowing with just a few keystrokes just how many people in a unit setting to deploy need immunizations provides the unit commander timely information to rectify the situation so that mission requirements can be met.
- The ability of Public Health technicians to daily monitor aggregated data regarding medical diagnoses can rapidly highlight the "spike" arising from the sudden occurrence of several cases each of a particular infectious disease (e.g. gastroenteritis) at three separate bases in theater. As a result, patients can be re-examined in a timely fashion and appropriate treatment and other countermeasures can be undertaken to reduce the spread of the outbreak. Further, such data can provide the epidemiologic evidence that could pinpoint the source of infection, such as a specific patient or food vendor.
- In peacetime, the vast storehouse of rich data that includes diagnoses and treatments rendered can be used in outcome studies to determine best practices. For example, through the "mining" of relevant data, a researcher could conduct a retrospective study of lower back pain, perhaps dividing the cases into samples of 10,000, each with a different treatment modality (medication vs. surgery, for example) and then compare outcomes to see if there is a statistical, and more importantly, a clinically significant difference in outcome.

## REVIEW OF CHAPTER GOALS

The goals of this chapter were:

1. To allow you to understand what a net-centric operational environment is all about:
    - Describe some of the benefits that makes it different from today's operations
2. To engage in a discussion on what types of problems our military is currently facing:
    - Address how the rigidity of a platform-centric environment can be made more agile by becoming net-centric
3. To provide an illustration of the Net-Centric Operational Context:
    - Describe each of the elements configured within this context
4. To promote the core aspects of sharing knowledge and assuring connectivity:
    - Address shared knowledge, collaboration, technical connectivity, distributed decision making, and network management principles
5. To describe knowledge management principles within the operational environment:
    - Describe the information, physical, social, cognitive domains and knowledge sharing mechanisms
6. To provide a case study example of a future net-centric operational environment:
    - Describe how a massive earthquake in Turkey is operationally supported and executed

## REFERENCES

Department of Defense Chairman of the Joint Chief of Staff (CJCS) Instruction 3710.01B (2001). *Requirements Generation System*. Retrieved from http://www.dtic.mil/doctrine/jel/cjcsd/cjcsi/3170_01b.pdf

GAO Report to the Chairman, Government Reform, House of Representatives, GAO-07-606T (2007, March). *DoD and VA Health Care Challenges Encountered by Injured Service members during their Recovery Process*.

Ghosh, S. et al., (2008, March). *VA/DoD Plan for Interagency Health Image Sharing of Essential Health Images*. VA/DoD Partnership in Response to Wounded, Ill, and Injured Senior Oversight Committee. Department of Defense Military Health System. (2004, May). *Theater Medical Information Program Capability Production Document (CPD), Increment 2*.

Infrastructure Management & Policy Development Workgroup, Health Information Exchange (2006, September). *Task Force to Study Electronic Records*.

Janczewski, M., & Ghosh, S. (2007). *Functional CONOPS for VA/DoD Joint Incentive Fund, Digital Imaging Initiative*. Veteran's Health Administration and Military Health System.

Office of the Assistant Secretary of Defense for Networks and Information Integration, Department of Defense Chief Information Officer (2008). *Department of Defense Information Management/ Information Technology Strategic Plan 2008-2009*.

Office of the Assistant Secretary of Defense for Networks and Information Integration, Department of Defense Chief Information Officer (2008). *Department of Defense Information Management/ Information Technology Strategic Plan 2008-2009*.

Young, G. J., (2000, June). *Transforming Government: The Revitalization of the Veterans Health Administration*. PricewaterhouseCoopers Endowment for the Business of Government.

# Section 3
# Configuring the Net-Centric Enterprise

*This section provides a number of specific topics and enabling technologies that encompass the layered concept of Net Centricity and helps to configure tomorrow's enterprise. There are six chapters within this section.*

- ***Chapter 8:*** *Use of Enterprise Architecture as a Net-Centric Discipline*
- ***Chapter 9:*** *Net-Centric Information Assurance Strategy*
- ***Chapter 10:*** *Adhering to Open Technology Standards*
- ***Chapter 11:*** *Service-Oriented Architecture and Net-Centric Principles*
- ***Chapter 12:*** *Transition to IPv6-Based Networks*
- ***Chapter 13:*** *Storage Strategy for the Distributed Enterprise*

# Chapter 8
# Use of Enterprise Architecture as a Net-Centric Discipline

## CHAPTER CONTENT

As you explore Chapter 8, it will cover the following topics:

- An Understanding of Enterprise Architecture
- Developing the Organization's EA
- EA Planning Guidelines
- Defining the EA Current and Future State
- Net-Centric Enterprise Architecture
- Integrated DoD EA Views
- Guidance for Net-Ready Key Performance Parameters

DOI: 10.4018/978-1-60566-854-3.ch008

## CHAPTER FOCUS

This chapter provides an understanding of enterprise architecture in action. We provide planning guidelines for developing organizational enterprise architecture and governance that is associated with it. Further details are then provided on developing an organization's architecture, and how we can define the enterprise's current and future state. We discuss net-centric enterprise architecture principles and discuss DoD integrated architecture views according to the current DoDAF that includes the System View, Operational View and Technical Standards View. The last section acknowledges that having enterprise architecture qualifies as a Net-Ready Key Performance Parameter, and then discusses how document an NR-KPP should be documented.

## AN UNDERSTANDING OF ENTERPRISE ARCHITECTURE

The practice of developing, implementing, and managing an Enterprise Architecture (EA) is recognized by both private and public organizations as critical to accomplishing organizational goals. Within the federal government, the topic of managing and planning business and technology resources was first addressed within the Information Technology Management Reform Act of 1996, which is better known as the Clinger-Cohen Act. The Clinger-Cohen Act mandates that federal agencies consider creating an enterprise architecture framework upon which to assess their information technology investments. The E-Government (E-Gov) Act of 2002 then gave broad EA oversight responsibilities to the Office of Management and Budget (OMB) which then placed EA mandates for each federal agency as part of their yearly federal budget submissions.

Successful enterprises use an EA to provide a clear and comprehensive picture of how the organization operates today, how it plans to operate in the future, and how it plans to transition to its future state. For any large organization, the preparation of an EA is critical to leveraging Information Technology (IT) in support of business objectives, particularly for organizational transformation initiatives. When used in conjunction with other important IT management controls such as portfolio-based capital planning and investment control practices, an EA can greatly increase the likelihood that an organization's business and IT environments will be effectively integrated so as to optimize the performance of the organization.

An EA helps an organization to recognize and balance the trade-offs between:

- satisfying immediate operational needs,
- and positioning the organization to achieve its longer-term strategic goals.

Here are a set of four precepts that help you understand the structure of an EA within the organization:

1. It is best recognize that EA addresses both business and technology needs.
2. Planning initiatives recognize that business planning drives the need for automation, which drives the need for information technology infrastructure.
3. In preparing the architecture for the enterprise, both the current state and the desired end state EA must be understood and documented.

4. There has to be an implementation plan put together to move forward from the current state to the target state the sequences the path to the future based on both business and technology elements.

## DEVELOPING THE ORGANIZATION'S EA

Setting up an EA for any government organization or large corporate firm is a significant undertaking. The development of a full set of EA artifacts is a large task and can be a difficult activity depending on the size and scope of the effort. The general consensus is to develop a core set of products that is guided by external legislative and regulatory requirements and then incrementally improve these products over time. Leadership within the organization needs to spend time figuring out how an EA program should be set up and how each of the EA deliverables should be created over time. The general guidance for an EA program within an organization would be to address the following issues:

- Align technology investments for systems and infrastructure to the agency's mission and goals
- Systematically engineer systems to integrate and interoperate within the internal and external environment
- Efficiently manage the implementation of emerging technologies
- Provide a comprehensive reference list of the information technology assets throughout the enterprise
- Provide ongoing knowledge management and information sharing to ensure that applications and communication mechanisms are kept current.

For a government agency, a documented EA needs to follow a fully compliant approach that encompasses all legal and regulatory requirements. These regulatory requirements may seem overwhelming, but it should be understood that while they all are important, not all can be implemented at once. An agency's implementation strategy is dependent upon business need, and resource availability. A successful practice would be for each agency to codify the following core set of elements in their EA requirements for both the current, or "As-Is" state and the future or "To-Be" state:

- **Business Processes** – Identify each of the key business processes that support the agency mission and objectives and codify it in a structured manner.
- **Information Flows** – Organize in a structured manner, information that is utilized by the agency, the information flows that indicate where it is needed and how it is being shared.
- **Applications** – Identify systems and applications that are resident within the Agency that automate and manage the business processes and information activities.
- **Data Descriptions and Relationships** – Organize common data elements that are used throughout the agency and identify how the data is created, maintained, accessed and stored.
- **Technology Infrastructure** – Organize and store in a visual manner, the agency technology infrastructure that includes the network and communications, and hardware and software of each system.
- **Technical Reference Model** – Provides a general framework on how technology is used throughout the agency, a strategy for technology refreshment and how new technology is inserted into existing programs.

- **System Standards Profiles** – Align each technology investment within an agency to make sure that common technology standards are being implemented throughout the organization.
- **Information Assurance** – Describe the information security controls and services within the EA products and deliverables.

## EA PLANNING GUIDELINES

As we plan for an EA within the organization, we need to recognize that this can be a slow and deliberate process.

Here are three major objectives to achieve as you set up an EA for your organization:

1. Obtain executive management support and the requisite resources for the EA Program,
2. Establish a formal organizational structure for the EA Program, and
3. Establish the program infrastructure that is required to support the EA process. This includes methods, plans, governance boards, technology tools, guiding principles, frameworks, and reference standards.

Advancing EA Maturity (S.Ghosh, et. al, 2005) provides the following steps to help prepare associated work products that become part of your EA artifacts and your organization's approach to move forward.

- **Determine the Scope and Objectives for EA Planning** – Work products include a definition of the scope of the enterprise and identification of the participating organizations - those that need to integrate and share processes and information.
- **Create a Vision Statement** – Work products include a published Vision Statement, briefings of the vision to management, and briefings of the vision to stakeholders - to establish expectations, create enthusiasm, and generate political support.
- **Obtain and Confirm Commitment and Funding** – A presentation work product is used to assure that executive management and key stakeholders understand the commitments needed to develop an EA.
- **Charter the EA Program** – This includes preparation, distribution, and review of the EA Program Charter. Activities resulting from an approved charter include the establishment of the EA Program, the EA Review Board, and the EA Management Team.
- **Adopt a Planning Methodology** – The EA planning methodology should address the business, data, applications, and technology elements of the enterprise. The methodology should also address the current state, target environment, and a sequencing plan. The methodology should be tailored to suit the needs of the organization.
- **Adopt an EA framework** – A framework is established to organize and structure the EA work products around business, data, applications, and technology perspectives. These can then be illustrated as views of the enterprise. The framework also identifies the critical elements that influence the EA, which includes strategic plans and oversight directives and the foundational elements that support it. This includes the organization's IT lifecycle processes and governance boards. Frameworks should also be established for each architectural view within the EA (e.g.

data, applications, and security) to structure and identify the specific work products required within each view.

- **Prepare EA Documentation** – An annotated outline for the entire EA documentation set should be developed to clarify what information is included in each major section of each work product. DoD organizations follow the DoDAF guidance as set by the DoD CIO and its prescribed set of EA views. A large organization may then decide to document, publish and display their documented products, online for users and stakeholders to review.

## DEFINING THE EA CURRENT AND FUTURE STATE

The development of an EA requires the agency to explicitly describe and document the current and desired relationships between the business process and information technology. By documenting the current "As-Is" state, the EA helps define the organization's principles and goals on such issues as interoperability with external systems, compliance with organization mandates, and traceability to end user needs. When the current state is defined, the EA then provides a strategy on how the agency can develop a roadmap that will support its current state and allow effective transition to the target environment. The transition process includes the agency's capital planning and investment control processes, agency EA planning processes and the agency systems life-cycle methodologies.

The target architecture that formulates the "To-Be" vision needs to have the agreement of the key stakeholders within the organization. A well-developed plan needs to be put in place that incorporates information from the strategy sessions held within the organization. A complete and institutionalized EA Program that codifies the target architecture consists of a sustainable governance framework and optimized business processes supported by appropriately configured data, applications, systems and technology infrastructures.

The development of a future or common "To-Be" vision for the enterprise has to be completed in the earliest phase of the EA development life cycle. A successful practice is to create a mission or vision statement that fully documents the vision of the organization stakeholders and aligns the vision to the agency core mission and goals. An architecture strategy document then articulates this agency vision into a Concept of Operations (CONOPS). The next steps are to examine the existing business definitions and processes and to analyze how the processes will change to accommodate the new "To-Be" state. The definitions and processes are then vetted against existing regulations such as the OMB Circulars, E-Gov and Clinger-Cohen Acts.

The process of tracing differences between the "As-Is" and the "To-Be" state is usually one of the most difficult steps in the entire EA life-cycle process. Agencies are expending a significant amount of time and resources in the activity of tracing enterprise requirements from the current to the target state.

The Department of Defense (DoD) Business Enterprise Architecture (BEA) has developed an architecture development methodology (BEA v.6.0, 2009) that provides details on how to take the existing set of architecture products and update it for the "To-Be" environment. The steps for this activity are the following:

- **Update EA Work Products** – Review and document required changes within each of the architecture views.
- **Integrate EA Views and Products** – Maintain linkages between each of the architecture views internal and external to the enterprise domain.

- **Verify EA Product Updates** – Conduct peer reviews and cross-review audits to determine if changes have been properly propagated.
- **Update Traceability to Requirements** – Requirements traceability should be codified in a repository that maintains the two-way linkages.
- **Update Business Scenarios** – Business scenarios used within the agency are updated based on change in architecture requirements.
- **Verify and Validate Architecture** – Analyze the "To-Be" model to ensure that changes administered by upcoming development programs have been incorporated.
- **Publish/Communicate to Stakeholders** – This activity covers tasks needed to easily disseminate the new architecture to all of the enterprise domains.

## NET-CENTRIC ENTERPRISE ARCHITECTURE

Now that we have talked about EA activities in general, it would be appropriate to discuss EA as an enabling activity for the net-centric transformation. The architectural foundation for the DoD Global Information Grid concept follows the net-centric model for integrating all DoD information systems. DoD guidance for net-centric operational warfare mandate compliance to the DoD EA framework and the current DoD information enterprise architecture.

So, let's further discuss the concept of "Net-Centric Enterprise Architecture". This term is starting to become of use within the military and civilian marketplace. In this book, we can define this term as follows:

**Net-Centric Enterprise Architecture** - is a large, distributed architecture with technology and service components that are available across and throughout the entire enterprise's line of business, products and solutions.

As our technology progresses, a Net-Centric Enterprise Architecture allows us to create an architectural mechanism that distributes components and/or services across an enterprise's information value chain using Internet technologies and common network protocols for supporting the distribution and processing of information services.

A Net-Centric Enterprise Architecture promotes the concept of an integrated architecture framework. Here we need to recognize that there is a hierarchy of architecture activities before the entire enterprise can be codified. The vision provided by the EA allows the development of consistent and appropriate systems across the enterprise with the ability to work together, collaborate, or integrate where and when required.

In this case, we also need to discuss two other architecture efforts that may be ongoing simultaneously within the organization.

- **Project-Level Architecture** – This defines the form and function of business and technology projects and the systems that they create within the context of the enterprise. It is expected that the project-level architecture will refine, conform to and work within the defined EA for the organization.
- **Application Architecture** – This defines the form and function of the applications that are developed to deliver the required functionality of the system that is defined by the project. It is expected that some of this architecture is defined by the organization EA and Project-level Architecture standards and guidelines to ensure best-practice and conformance to the overall architecture.

*Table 1. Description of DoDAF EA Views*

| DoD EA Views | Description |
|---|---|
| All View (AV) | The All Views (AV) products are intended to provide an overall perspective of the operational enterprise, and address overarching concerns like Concept of Operation (CONOPS) and critical mission objectives and strategies, as well as an integrated dictionary of architecturally significant terms. |
| Operational View (OV) | The Operational Views (OV) focus on the externally visible structure and behavior of the enterprise. This view describes operational nodes and their relationships, and identifies dependencies reflecting mission requirements, thus providing an overall context for enterprise definition and evolution. |
| System View (SV) | The Systems Views (SV) focus on the realization of internal structure and behavior of the enterprise, a review of each system module and interfaces, a rigorous allocation of system requirements in correlation with the Operational View, and delineate both logical and physical system elements and interfaces. |
| Technical Standards View (TV) | The Technical Standards View (TV) provides system, data and operational standards that constrain the operational architecture of the enterprise and needs to be reflected in both the current and future states of the enterprise |

The DoD is currently using net-centric enterprise architecture principles within the Defense Information Enterprise Architecture (DIEA). This is discussed (DIEA v1.0, 2008) as providing a common foundation to support the DoD transformation to net-centric operations. In this case, the Defense Information Enterprise comprises of information, information resources, assets, and processes required to achieve an information advantage and share information across the military and its mission partners. DIEA target EA goals are to enable agile, collaborative net-centric operations. In today's information environment the DIEA ensures for example, that a persistently-connected Internet Protocol (IP) exist based on boundaries of the Global Information Grid (GIG).

## INTEGRATED DoD EA VIEWS

For large organizations, it is best that EA is presented in an integrated manner that encompasses all aspects of the business, technology, people and processes. Based on net-centric enterprise architecture principles, this section focuses specifically on DoD EA products. Commercial organizations may choose to use other EA frameworks and create artifacts that are suited for their organization.

For DoD component organizations, it is expected that an integrated architecture approach is adopted. Integrated architectures provide an illustrative representation of the DoD program's processes, systems, and system interfaces. According to existing guidelines of the DoD Architecture Framework (DoDAF), an integrated architecture is comprised of the following primary views. As the DoDAF is revised for the future, these architecture views may become enhanced to include other views or perspectives.

Table 1 provides definitions of each of the views that are currently in use based on DoDAF Version 1.5 (DoDAF v1.5 Volume 1, 2007). We understand that DoDAF Version 2.0 is being released by the DoD, and it has a number of views and viewpoints that are additional artifacts. However, at this time, as has been practiced by the large defense community, the EA views can be described as to the following set.

DoDAF specifies a common approach for describing, comparing and presenting enterprise-wide DoD component architecture. This information is stated based on the current DoDAF Version 1.5 guidance.

*Figure 1. Integrated DoDAF View Relationships*

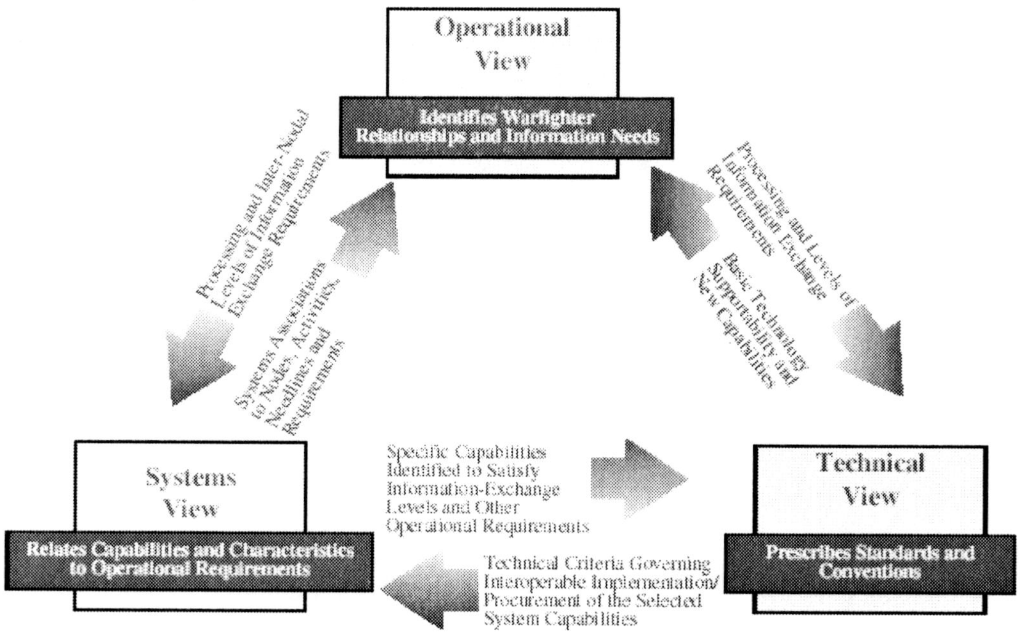


This approach may change as the DoDAF is revised in the future. At this time, it describes a set of 26 architecture product views to ensure uniformity and standardization in documenting defense-wide architectural concepts.

Figure 1 is excerpted from the DoD documentation (DoDAF v1.5 Volume 1, 2007) and illustrates the architectural relationships between the integrated views of OV, SV and TV.

## All View (AV)

All View (AV) products provide the overarching description of the entire architecture and define the scope and context of the architecture. The AV products are defined as:

- **AV-1:** Overview and Summary Information
- **AV-2:** Integrated Dictionary

## Operational View (OV)

The OV products provide descriptions of the tasks and activities, operational elements, and information exchanges required to accomplish DoD missions. The OV provides textual and graphical representations of operational nodes and elements, assigned tasks and activities, and information flows between nodes. It defines the type of information exchanged, the frequency of exchanges, the tasks and activities supported by these exchanges and the nature of the exchanges. The OV products are defined as:

- **OV-1:** High Level Operational Concept Graphic
- **OV-2:** Operational Node Connectivity Description
- **OV-3:** Operational Information Exchange Matrix
- **OV-4:** Organizational Relationships Chart
- **OV-5:** Operational Activity Model
- **OV-6a:** Operational Rules Model
- **OV-6b:** Operational State Transition Description
- **OV-6c:** Operational Event-Trace Description
- **OV-7:** Logical Data Model

## Systems View (SV)

The SV products provide graphical and textual descriptions of systems and system interconnections that provide or support DoD functions. Interconnections between systems defined in the OV are described in the SVs. The SV products are:

- **SV-1:** System Interface Description
- **SV-2:** Systems Communications Description
- **SV-3:** Systems-Systems Matrix
- **SV-4:** Systems Functionality Description
- **SV-5:** Operational Activity to Systems Functionality Traceability Matrix
- **SV-6:** Systems Data Exchange Matrix
- **SV-7:** Systems Performance Parameters Matrix
- **SV-8:** Systems Evolution Description
- **SV-9:** Systems Technology Forecast
- **SV-10a:** Systems Rules Model
- **SV-10b:** Systems State Transition Description
- **SV-10c:** Systems Event-Trace Description
- **SV-11:** Physical Schema

## Technical Standards View (TV)

The TV products define technical standards, implementation conventions, business rules and criteria that govern the architecture. The TV products are as follows:

- **TV-1:** Technical Standards Profile
- **TV-2:** Technical Standards Forecast
- **SV-TV Bridge:** this allows direct correlation of technical standards to a DoD component system view

Currently, the DoD Information Systems Agency (DISA) has purview of DoD technical standards for the entire agency. They have created a DoD IT Standards Registry (DISR) which is a repository of technical standards in use within component organizations. DoD organizations are urged to register within the DISR, and create technical standards profiles that then serve to bring greater interoperability to the defense and military organizations.

## GUIDANCE FOR NET-READY KEY PERFORMANCE PARAMETERS

Note that we have discussed net-centric enterprise architecture principles, this section provides further guidance based on DoD doctrines and instructions. DoD interoperability guidance provided within in DOD Instruction 4630.8 (DoDI 4630.8, 2004) provides a definition of Net-Ready Key Performance Parameters (NR-KPPs). This section highlights that for DoD component organizations, having an integrated EA for their enterprise qualifies as a Net-Ready Key Performance Parameter (NR-KPP). This is also a key interoperability parameter as discussed within further guidance given by DoD joint instructions (CJCSI 6212.01E, 2008).

### Definition of a Net-Ready Key Performance Parameter (NR-KPP)

The definition for a NR-KPP may be stated as follows:

- An NR-KPP assesses information needs, information timeliness, information assurance, and net-ready attributes required for both the technical exchange of information and the end-to-end operational effectiveness of that exchange.
- An NR-KPP consists of verifiable performance measures and associated metrics required to evaluate the timely, accurate, and complete exchange and use of information to satisfy information needs for a given capability.

As part of the DoD acquisition process, DoD programs and systems need to document NR-KPPs within their milestone documents as they are developed within the program or system development life cycle.

For a representative DoD program that has agency-wide visibility, a documented NR-KPP is comprised of the following elements:

- **Compliance with the Net-Centric Operations and Warfare (NCOW) Reference Model (RM):** The NCOW RM describes the activities required to establish, use, operate, and manage the net-centric enterprise information environment. This environment includes the generic user-interface, the intelligent-assistant capabilities, the net-centric service capabilities such as core services, Community of Interest (COI) services, and environment control services, and the enterprise management components. It also describes a selected set of key technical standards that will be needed as the NCOW capabilities of the GIG are realized.
- **Compliance with Applicable GIG Key Information Parameters:** GIG information parameters provide a net-centric oriented approach for managing interoperability across the GIG based on the configuration control of key interfaces.
    - A GIG key information parameter is the set of documentation produced as a result of interface analysis, which designates an interface as key

*Table 2. Architecture Products Required for Net Readiness*

| **Architecture Products Required to Assess Information Exchange and Use** | | |
|---|---|---|
| **Framework Product** | **Framework Product Name** | **General Description** |
| **AV-1** | All View Overview and Summary | Provide a text description of the enterprise and associated programs, systems and facilities |
| **AV-2** | Integrated Dictionary | A dictionary of terms, glossary of definitions |
| **OV-1** | High-Level Operational Concept | Scope, purpose, intended users, environment depicted, analytical findings |
| **OV-2** | Operational Node Connectivity Description | Operational Nodes, operational activities performed at each node, connectivity and information exchange between nodes |
| **OV-3** | Information Exchange Requirements matrix | A spreadsheet of information requirements that is being shared between entities |
| **OV-4** | Organizational Relationships Chart | Organizational, role, or other relationships among organizations |
| **OV-5** | Operational Activity Model | Operational Activities, relationships among activities, inputs and outputs. Overlays can show cost, performing nodes, or other pertinent information. |
| **OV-6c** | Operational Event-Trace Description | One of three products used to describe operational activity sequence and timing – traces actions in a scenario or sequence of events and specifies timing of events |
| **OV-7** | Logical Operational Data Model | A diagram that provides contextual understanding of data items of the enterprise |
| **SV-1** | Systems Interface Description | A comprehensive diagram of systems and how they are interrelated within the enterprise |
| **SV-2** | Systems Communication Description | A diagram that captures communications mechanisms and protocols for system interfaces |
| **SV-4** | Systems Functionality Description | Functions performed by systems and the information flow among system functions |
| **SV-5** | Operational Activity to System Function Traceability Matrix | Mapping of systems back to operational capabilities or of system functions back to operational activities |
| **SV-6** | Systems Data Exchange Matrix | Provides details of systems data being exchanged between systems |
| **SV-11** | Systems Physical Schema | Provides an physical understanding of data schema of the systems and how they inter-relate |
| **TV-1** | Technical Standards Profile | Extraction of standards that apply to the given architecture |
| **TV-2** | Technical Standards Forecast | Captures the evolution of technical standards that are in use throughout the enterprise |

- The interface parameters are analyzed to understand its architectural, interoperability, test and configuration management characteristics
- The interface profile consists of refined operational and systems view products, Interface Control Document Specifications, Engineering Management Plan, Configuration Management Plan, TV-1 with SV-TV Bridge, and procedures for standards conformance and interoperability testing.
- Relevant GIG information parameters for a given capability are documented within program milestone documents.
- The Chairman of the Joint Chiefs of Staff will continue to refine information parameters within the GIG. Technical standards for GIG interface parameters would be maintained in the DISR.

- **Maintain Compliance with DoD Information Assurance Requirements:** Verification of compliance with DoD IA requirements as specified within the latest DoD instructions, and DIACAP guidance.
- **Supporting Integrated Architecture Products:** The following integrated architecture products described as a minimum, be incorporated in the NR-KPP and used to assess information exchange and use for a given capability:

### Supporting EA Products for NR-KPP Compliance

Even though, it is best to have a full set of integrated EA products, the DoD instructions have figured out that a certain set of architecture products provide more direct benefit to assess a system's net-centric capabilities.

Table 2 provides an understanding of a minimum set of EA products based on CJCSI guidance (CJCSI 6212.01E, 2008) that is necessary to determine the tactical functions of a system within the net-centric environment.

## REVIEW OF CHAPTER GOALS

The goals of this chapter were to address:

1. How can we adopt fundamental principles and objectives of enterprise architecture planning and methodology:
   - We hold true to all of enterprise architecture planning efforts to address both business and technology
2. What elements should an organization's enterprise architecture address:
   - This would include business processes, information flows, applications and technology infrastructure
3. How to go ahead and document the organization's current state and define the upcoming target state:
   - Document both the "as-is" and the "to-be" environment
4. How can an integrated set of architecture views can be depicted:
   - Use the DoDAF to provide three-sided model of system, operational and technical standards views
5. What are the names and description of each of the DOD architecture views:
   - A listing and description of each of the enterprise architecture products
6. For the definition of a Net-Ready Key Performance Parameter:
   - Provide a documented set of evidence to comply with the net-centric readiness for DoD programs

## REFERENCES

Bass, T., Mabry, R., (2004, April). *Enterprise Architecture Reference Models: A Shared Vision for Service-Oriented Architectures*, (Version 0.81). IEEE Milcom 2004.f

Chairman of the Joint Chiefs of Staff Instruction 6212.01E (2008, December). *Interoperability and Supportability of Information Technology and National Security Systems*. Washington, DC: Author.

*Commonwealth of Massachusetts Enterprise Information Technology Architecture, Enterprise Technical Reference Model – Version 3.5,* (2005, September 21).

Department of Defense. (2007). *DoD Architecture Framework Version 1.5 - Volume I: Definitions and Guidelines*. Washington, DC: Author.

Department of Defense Business Transformation Agency. (2009). *BEA Architecture Product Guide Version 6.0*. Retrieved from http://www.bta.mil/products/bea.html

Department of Defense Instruction 4630.8 (2004, June). *Procedures for Interoperability and Supportability of Information Technology (IT) and National Security Systems (NSS)*. Washington, DC: Author.

*Executive Office of the President, Federal Enterprise Architecture Program EA Assessment Framework 2.0,* (2005, December).

Executive Office of the President (2007, October). *FEA Consolidated Reference Model Document,* (Version 2.3).

Ghosh, S., Miller, C., Przysucha, J., & Tiemann, M., (2005, January). *Advancing Enterprise Architecture Maturity, Version 2.0*. Industry Advisory Council White Paper.

Office of the Assistant Secretary of Defense for Networks and Information Integration, Department of Defense Chief Information Officer (2008, April). *Defense Information Enterprise Architecture,* (Version 1.0).

Spewak, S. H. (2002). *Enterprise Architecture Planning: Developing a Blueprint for Data, Applications, and Technology*, (2nd Ed.). Chichester, UK: John Wiley & Sons.

# Chapter 9
# Net-Centric Information:
## Assurance Strategy

## CHAPTER CONTENT

As you explore Chapter 9, it will cover the following topics:

- Information Assurance Definitions
- Information Assurance Certification & Accreditation
- Use of Common Criteria Standards
- Net-Centric Information Assurance Vision
- Information Assurance for Net-Centric Operations

## CHAPTER FOCUS

This chapter now shifts to describing Information Assurance and how it is a crucial topic that aids the process of Net Centricity. We start with a set of information assurance definitions that have been stated in the language within DoD guidance documents. We then define common information assurance prin-

DOI: 10.4018/978-1-60566-854-3.ch009

ciples and define terms such as Defense-in-Depth. We then discuss the certification and accreditation process that has been standardized throughout the DoD to ensure that security measures are in place. The latest DoD process is DIACAP, and it involves implementing a set of information assurance steps. We then introduce Common Criteria standards, an international framework that allows government and industry to work together on a common method for implementing security measures. Then we discuss net-centric information assurance goals and how information assurance aspects are adopted for net-centric operations.

## INFORMATION ASSURANCE DEFINITIONS

Since Information Assurance (IA) is a discipline that is complex and not well understood, this section tries to introduce the subject matter by providing definitions of fundamental terms. There are a large number of defense and military guidance documents, written as DoD Directives and DoD Instructions that cover the fundamentals of Information Assurance. Information assurance definitions are also used by other government and civilian entities, and the concepts are similar in nature.

The following set of definitions has been obtained based on DoD Directive 8500.1, titled, "Information Assurance" (DoD 8500.1E, 2007). This information assurance directive is used by U.S. defense organizations that include DoD Components, Armed Force Services, and associated federal government agencies and civilian organizations.

### Definition of Information Assurance

Here is a definition of Information Assurance based on the DoD Information Assurance Certification and Accreditation Process (DIACAP) documentation and DoD net-centric guidelines.

- Information Assurance (IA) is the discipline that provides methods for managing the risks of information assets
- Information Assurance is defined as measures taken by an organization to protect and defend information and information systems to ensure the following principles:
    - Availability
    - Integrity
    - Authentication
    - Confidentiality
    - Non-Repudiation
- This includes providing for the restoration of information systems by incorporating protection, detection and reaction capabilities
- Information Assurance protects data whether the data is in storage, processing, or transit, and whether the data is threatened by malice or by accident

### Definition of Information Assurance Principles

The following five information assurance principles, in total provide the breadth and coverage of information assurance and provide a fuller understanding of the discipline.

- **Availability** – Availability of data is based on the premise of timely and reliable access to data and information services for authorized users
- **Integrity** – Refers to the quality of the information and information system; it may reflect the logical correctness and reliability of the operating system, the completeness of the hardware and software that is implementing the protection mechanisms; and the consistency of the data structures and occurrence of the stored data integrated within the information system.
- **Authentication** – This property is described as a security measure that is designed to establish the validity of a data transmission, the message, the originator of the message or the means to verify the individual's authorization to view, access or receive specific information sets.
- **Confidentiality** – A measure that allows you to assure that information is not disclosed to any unauthorized entities or information processes
- **Non-repudiation** – This allows you to have assurance that the sender of the data is provided with the proof of delivery, and the recipient of the data is provided with the proof of the sender's identity. This allows neither party to later deny that had not processed the data

Other definitions of common terms that stem from these five principles are:

- **Certification and Accreditation (C&A)** – This is a standard information assurance process created by the DoD for information and information systems to identify and document information security requirements, provide security solutions, and manage the security implementation of DoD information systems
- **Authorized User** – Any individual or entity that has been appropriately cleared, so that they have a requirement to access a DoD information system. Their objective is to perform and assist in a lawful and authorized government function within an information system.
- **Encryption** - This is the process of protecting the confidentiality of information by transforming the information based on an algorithm that makes it unreadable to anyone except those possessing special knowledge, usually referred to as a key. The result of this process gives you encrypted information. The word encryption also is aligned with the reverse process of decryption that makes encrypted information readable again. There are a number of software techniques and algorithms for encryption and decryption.

## Defining Defense-in-Depth

Defense in Depth is a practical strategy for Information Assurance developed by the DoD to establish an adequate security posture for participating organizations. The strategy recommends that there should be a balance between the protection capability provided by information assurance principles, and the information system cost, performance and operations. Defense-in-Depth is a best practices strategy that relies on the intelligent application of assurance techniques and technologies. The recommendation is for a shared-risk environment that allows for shared mitigation of end-user solutions. This mitigation relies in the layering of information assurance solutions within and among information system assets. Figure 1 provides a broad understanding of Defense-in-Depth as a strategy.

An important principle of the Defense in Depth strategy is that achieving Information Assurance requires a balanced focus on three primary elements: People, Technology and Operations.

*Figure 1. Defense-in-Depth Strategy*

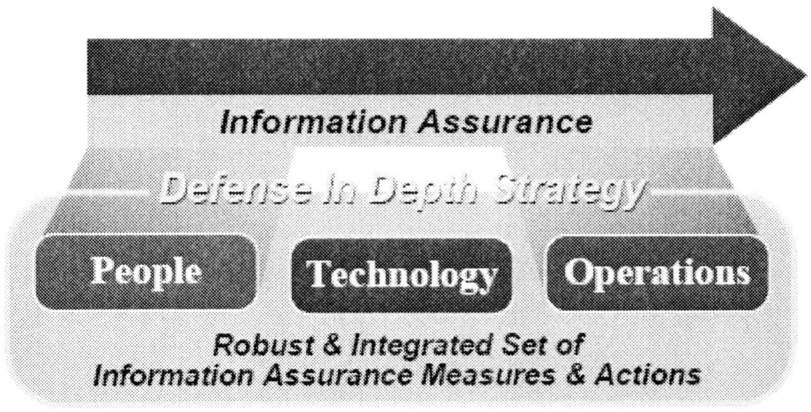


- **People** – Achieving Information Assurance begins with a senior level management commitment (typically at the Chief Information Officer level) based on a clear understanding of the perceived threat. This must be followed through with effective Information Assurance policies and procedures, assignment of roles and responsibilities, commitment of resources, training of critical personnel (e.g. users and system administrators), and personal accountability.
- **Technology** - Today, a wide range of technologies are available for providing Information Assurance services and for detecting intrusions. To insure that the right technologies are procured and deployed, an organization should establish effective policy and processes for technology acquisition. These should include: security policy, Information Assurance principles, system level Information Assurance architectures and standards, criteria for needed Information Assurance products, acquisition of products that have been validated by a reputable third party, configuration guidance, and processes for assessing the risk of the integrated systems.
- **Operations** - The operations leg focuses on all the activities required to sustain an organization's security posture on a day to day basis. These include:
  a) Maintaining visible and up to date system security policy
  b) Certifying and accrediting changes to the Information Technology baseline. The C&A processes should provide the data to support "Risk Management" based decisions. These processes should also acknowledge that a "risk accepted by one is a risk shared by many" in an interconnected environment.
  c) Managing the security posture of the Information Assurance technology (e.g. installing security patches and virus updates, maintaining access control lists)
  d) Providing key management services and protecting this lucrative infrastructure
  e) Performing system security assessments (e.g. vulnerability scanners, RED teams) to assess the continued "Security Readiness"
  f) Monitoring and reacting to current threats
  g) Attack sensing, warning, and response
  h) Recovery and reconstitution

## Defining Enclave

Within a distributed environment as is common with today's and tomorrow's enterprise, it is important to note the strict definition of a security enclave. An enclave is typically characterized as a collection of computing environments connected by one or more internal networks under the control of a single authority and security policy, including personnel and physical security.

Enclaves always assume the highest mission assurance category and security classification of the information system and applications. It also includes the outsourced IT-based processes they support, and derive their security needs from those systems. Enclaves provide a set of standard information assurance capabilities such as boundary defense, incident detection and response, and key management, and also deliver common user applications for the desktop and electronic mail. Enclaves may be specific to an organization or a mission. The computing environments may be organized by physical proximity or could be independent of location.

Examples of enclaves include Local Area Networks (LANs) and the applications they host within their backbone and their access to local data processing centers.

## INFORMATION ASSURANCE CERTIFICATION AND ACCREDITATION

DoD Information Assurance Certification and Accreditation Process (DIACAP) is currently the standardized methodology for evaluating the security posture of Department of Defense (DoD) information systems.

It is DoD policy that the Department of Defense will certify information systems through an enterprise process for identifying, implementing and management information assurance capabilities and services. IA capabilities and services are expressed as IA controls as defined in the DoD Instruction 8500.2, Information Assurance Implementation (DoDI 8500.2, 2003).

The DIACAP, as compared with the previously implemented DITSCAP, approaches the C&A process with a lifecycle and enterprise focus, encouraging and facilitating the implementation of C&A early in lifecycle requirements. This approach enables the early engagement of both IA personnel and other key stakeholders such as program managers, systems engineers and developers.

DIACAP enables the stakeholders to link requirements to appropriate IA controls. This includes both system and operational environment that are specific early in the lifecycle. This linkage injects C&A into the iterative development process, thus providing more accurate traceability between implementation and system risk. Figure 2 provides a revolving process, which starts with initiation and ends with decommission of a system.

1) **Initiate and Plan IA C&A** - This activity includes registering the system with the governing DoD Component IA program, assigning IA controls based on Mission Assurance Category (MAC) and Confidentiality Level (CL), identifying the DIACAP Team for the IS, and initiating the information system's defense posture.
2) **Implement and Validate Assigned IA Controls** - This activity includes executing the information posture, conducting validation activities, preparing the IT security program milestone, and compiling the validation results in the DIACAP scorecard.

*Figure 2. A Step-Wise Set of DIACAP Activities*

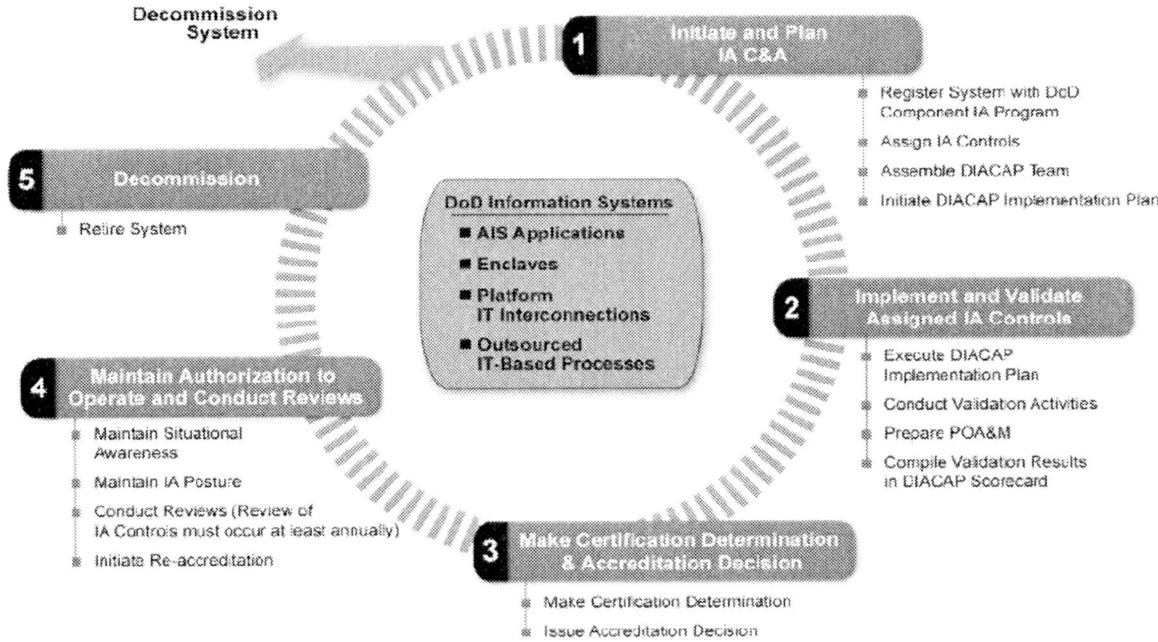


3) **Make Certification Determination and Accreditation Decision** – A Certification Authority (CA) makes certification determinations. A CA representative is an active member of the DIACAP Team from inception and continuously assesses and guides the quality and completeness of DIACAP activities and tasks and the resulting artifacts. The Certification considers the overall reliability and viability of the DoD information system plus the acceptability of the implementation and performance of IA mechanisms or safeguards inherent in the system.
4) **Maintain Authorization to Operate and Conduct Reviews** – To receive an Authority TO Operate (ATO) on a continued basis is contingent on the sustainment of an acceptable IA posture. The DoD information assurance management has primary responsibility for maintaining situational awareness and initiating actions to improve or restore IA posture.
5) **Decommission** - When a DoD information system is removed from operation, a number of DIACAP-related actions are required. Prior to decommissioning, any inheritance relationships should be reviewed and assessed for impact. Once the system has been decommissioned, DIACAP activity, System Life Cycle Phase should be updated to reflect the information system's decommissioned status.

## USE OF COMMON CRITERIA STANDARDS

We now provide information regarding Common Criteria, and international set of security and information assurance standards that has formed a consensus across the global community. Common Criteria standards provide a framework by which computer system users can specify their security requirements, vendors can then implement and/or make claims about the security attributes of their products, and testing laboratories can evaluate the products to determine if they actually meet the claims. In other words, Common Criteria provides assurance that the process of specification, implementation and evaluation of a computer security product has been conducted in a rigorous and standard manner.

This multi-part standard, the Common Criteria (CC), is meant to be used as the basis for evaluation of security properties of IT products. By establishing such a common criteria base, the results of an IT security evaluation may be meaningful to a wider audience.

The CC permits comparability between the results of independent security evaluations. The CC does so by providing a common set of requirements for the security functionality of IT products and for assurance measures applied to these IT products during a security evaluation. These IT products may be implemented in hardware, firmware or software.

The evaluation process establishes a level of confidence that the security functionality of these IT products and the assurance measures applied to these IT products meet these requirements. The evaluation results may help consumers to determine whether these IT products fulfill their security needs.

The CC is useful as a guide for the development, evaluation and/or procurement of IT products with security functionality.

- Products can be evaluated by competent and independent licensed laboratories so as to determine the fulfillment of particular security properties, to a certain extent or assurance;
- Supporting documents, are used within the Common Criteria certification process to define how the criteria and evaluation methods are applied when certifying specific technologies;
- The certification of the security properties of an evaluated product can be issued by a number of Certificate Authorizing Schemes, with this certification being based on the result of their evaluation

## General Model for Security and Protection of Assets

The field of security is concerned with the protection of assets. Assets are entities that someone places value upon. But given that value is highly subjective, almost anything can be termed an asset.

Safeguarding assets of interest is the responsibility of owners who place value on those assets. Actual or presumed threat agents may also place value on the assets and seek to abuse assets in a manner contrary to the interests of the owner. Examples of threat agents include hackers, malicious users, non-malicious users, computer processes and accidents.

The owners of the assets will perceive such threats as potential for impairment of the assets such that the value of the assets to the owners would be reduced. Security-specific impairment commonly includes, but is not limited to: loss of asset confidentiality, loss of asset integrity and loss of asset availability.

These threats therefore give rise to risks to the assets, based on the likelihood of a threat being realized and the impact on the assets when that threat is realized. Subsequently countermeasures are imposed to reduce the risks to assets. These countermeasures may consist of IT countermeasures (such as firewalls and smart cards) and non-IT countermeasures.

## Evaluation Assurance Levels (EALs)

This is the numerical rating describing the depth and rigor of a Common Criteria evaluation. Each EAL corresponds to a package of security assurance requirements which covers the complete development of a product, with a given level of strictness.

There are seven EAL ratings, and they are widely used for rating vendor equipment, hardware and software, and communications infrastructure.

### EAL1: Functionally Tested

EAL1 is applicable where some confidence in correct operation is required, but the threats to security are not viewed as serious. It will be of value where independent assurance is required to support the contention that due care has been exercised with respect to the protection of personal or similar information. EAL1 provides an evaluation of the TOE (Target of Evaluation) as made available to the customer, including independent testing against a specification and an examination of the guidance documentation provided. It is intended that an EAL1 evaluation could be successfully conducted without assistance from the developer of the TOE, and for minimal cost. An evaluation at this level should provide evidence that the TOE functions in a manner consistent with its documentation, and that it provides useful protection against identified threats.

### EAL2: Structurally Tested

EAL2 requires the cooperation of the developer in terms of the delivery of design information and test results, but should not demand more effort on the part of the developer than is consistent with good commercial practice. As such it should not require a substantially increased investment of cost or time. EAL2 is therefore applicable in those circumstances where developers or users require a low to moderate level of independently assured security in the absence of ready availability of the complete development record. Such a situation may arise when securing legacy systems.

### EAL3: Methodically Tested and Checked

EAL3 permits a conscientious developer to gain maximum assurance from positive security engineering at the design stage without substantial alteration of existing sound development practices. EAL3 is applicable in those circumstances where developers or users require a moderate level of independently assured security, and require a thorough investigation of the TOE and its development without substantial re-engineering.

### EAL4: Methodically Designed, Tested, and Reviewed

EAL4 permits a developer to gain maximum assurance from positive security engineering based on good commercial development practices which, though rigorous, do not require substantial specialist knowledge, skills, and other resources. EAL4 is the highest level at which it is likely to be economically feasible to retrofit to an existing product line. EAL4 is therefore applicable in those circumstances where developers or users require a moderate to high level of independently assured security in conventional commodity TOEs and are prepared to incur additional security-specific engineering costs.

*Figure 3. DoD Vision for implementing IA Goals*

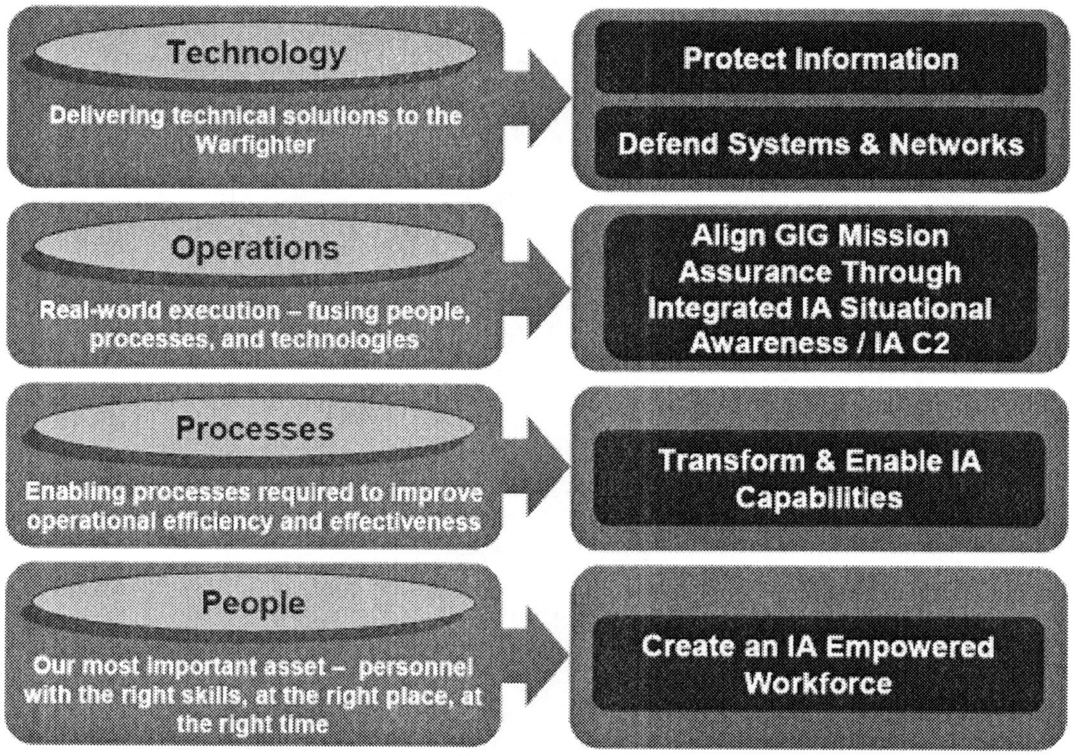


## EAL5: Semi-Formally Designed and Tested

EAL5 permits a developer to gain maximum assurance from security engineering based upon rigorous commercial development practices supported by moderate application of specialist security engineering techniques. Such a TOE will probably be designed and developed with the intent of achieving EAL5 assurance. It is likely that the additional costs attributable to the EAL5 requirements, relative to rigorous development without the application of specialized techniques, will not be large. EAL5 is therefore applicable in those circumstances where developers or users require a high level of independently assured security in a planned development and require a rigorous development approach without incurring unreasonable costs attributable to specialist security engineering techniques.

## EAL6: Semi-formally Verified Design and Tested

EAL6 permits developers to gain high assurance from application of security engineering techniques to a rigorous development environment in order to produce a premium TOE for protecting high value assets against significant risks. EAL6 is therefore applicable to the development of security TOEs for application in high risk situations where the value of the protected assets justifies the additional costs.

### EAL7: Formally Verified Design and Tested

EAL7 is applicable to the development of security TOEs for application in extremely high risk situations and/or where the high value of the assets justifies the higher costs. Practical application of EAL7 is currently limited to TOEs with tightly focused security functionality that is amenable to extensive formal analysis.

## NET-CENTRIC INFORMATION ASSURANCE VISION

According to the DoD, the current information assurance mission as stated within the DoD Information Assurance Strategic Plan (DoD IA Strategic Plan, 2008) is to accomplish the following:

*"Assure the Department's information, information systems and information infrastructure to support the Department's transformation to Network and Data-Centric Operations and Warfare and Global Information Grid (GIG) Mission Assurance."*

To realize this vision, each vision element has been directed to correspond with information assurance implementation goals. Figure 3 provides the illustration of this vision of implementing IA goals.

- **Goal 1** – Protect Information to safeguard data as it is being collected, analyzed, processed and disseminated wherever to ensure all information has a level of trust commensurate with mission needs.
- **Goal 2** – Defend Systems and Networks by recognizing, reacting to and responding to threats, vulnerabilities and deficiencies and by ensuring all systems and networks are capable of self-defense.
- **Goal 3** – Align GIG Mission Assurance Through Integrated IA Situational Awareness/IA Command and Control (C2) integrating the User-Defined Operational Picture (UDOP) synchronized with NetOps and emerging GIG Common Operating Picture (COP) programs to provide decision makers and network operators at all levels the capabilities for conducting IA/CND operations in Net-Centric Warfare (NCW).
- **Goal 4** – Transform and Enable IA Capabilities through innovation and experimentation, leveraging emerging technologies, operationalizing IA best practices and refining foundational processes to improve cycle time, reduce risk exposure and increase return on investments.
- **Goal 5** – Create an IA Empowered Workforce that is well equipped to support the changing demands of the IA/IT enterprise.

## INFORMATION ASSURANCE FOR NET-CENTRIC OPERATIONS

In Joint Net-Centric Operations, network information must be protected from attack and against unauthorized access, and yet remain continuously accessible by authorized users. Information assurance provides the Joint Forces with assured mission management, assured information sharing, confidentiality, and integrity/non-repudiation capabilities in support of enterprise network information and services during the conduct of joint actions.

*Table 1. Marching to the Target State for Information Assurance within the DoD*

| Current State | Interim State | Target State |
|---|---|---|
| • Non-Interoperable Point Solutions<br>• Build Foundation for Future | • Global unique digital identity and IA credentials for all IT components & complex systems<br>• Account for individuals, organizational users, COIs | • IA Privileges are dynamically adjusted to reflect IA posture of the GIG or entity |

As a critical component of the Net-Centric Operational Environment, IA will protect information from the time of its generation at network nodes, through its storage, processing, cataloging, and until its distribution. The distribution may occur via posting, smart push, or information pull strategies to individuals and groups of decision makers.

Non-repudiation provides a timely, highly accurate ability to verify the identity of both the sender and the receiver of the information and also that information's authenticity. In the upcoming future, information providers, processors, and consumers are targets for spoofing, masquerading, disclosure, data modification, system behavior modification, and/or denial-of-service attacks. Similarly, information transactions between each of those nodes may be targeted for eavesdropping, service denial, modification of data in-transit, mis-routing of information, and traffic analysis-types of attacks. This net-centric environment must defend against all of these elements. Since new threats are emerging even as the old ones continue to evolve, the IA strategy must be proactive in striving to understand how the NCOE's various facets, including its newest and emerging capabilities, could be exposed to disruption or manipulation.

## Necessary IA Capabilities

For net-centric operations, here are a set of six broad categories of IA capabilities that are necessary for upcoming operations within governmental organizations and also within large commercial organizations.

- **Assured Information Sharing**—providing the JTF with the means to positively identify and authenticate entities on the network, share information between authorized entities, as well as form and manage dynamic communities of interest to support collaboration.
- **Highly Available Enterprise**—ensuring that computing and communication resources, net-centric services, and information are available to support the JTF. This capability provides an appropriate level of protection in the presence of attacks, degradation in bandwidth quality/quantity, and service availability, while maintaining key performance parameters.
- **Confidentiality**—enables the ability to validate and assure that network systems or information is provided to the correct party and not disclosed to unauthorized sources. Confidentiality also ensures that only trusted parties participate in the communication exchange.
- **Defend the GIG**—requiring the tools, processes, and means to detect unauthorized activity that may affect information sharing and collaboration. It also provides a means to maintain GIG services and access to information while the NCOE is under attack, and to prevent future attacks
- **Integrity and Non-Repudiation**—enabling technical capabilities that provide for storing, sharing, exchanging, and processing information with the assurance that it is correct and valid. This directly enables the JTF's abilities to share situational awareness and to understand information

with confidence, knowing that what is communicated is what is intended. It also provides users with abilities to conduct collaborative decision-making and planning and to operate interdependently, confident that communication exchanges can be verified.

- **Assured Mission Management**—the ability to coordinate and de-conflict system configuration and resource changes, mission priority changes, and cyber-attack responses. It includes the ability to assign, prioritize, modify and revoke user-roles and system-roles, access rights, COI membership, and resources in a coordinated fashion using privileges, cryptographic keys, and IA configurations. Additionally, it ensures that attack responses do not adversely affect mission priorities. It must be automated, timely, and capable of reacting to changing priorities while under normal, degraded, and disconnected conditions.

## Upcoming IA Challenges

Table 1 provides a glimpse into the different state of information assurance activities and where the military and DoD intends to go in the future. Within the information assurance realm, this allows us to evaluate identity, IA credentials, and IA posture.

The following is a list that has been put together by DoD organizations that reflects the upcoming changes and expected challenges that is within the realm of information assurance.

- **Identity for Simple Entities**
    - These are for individuals, and leverages DoD PKI with identified improvements
    - These are HW/SW components that includes weapons systems, computing and communications infrastructure devices, user equipment, and software objects
- **Identity for Complex Entities**
    - Organizations
    - Information Systems and Networks
- **Full Spectrum Security Credentials and Privileges for Simple and Complex Entities**
    - Voice, Video, Data, Imaging
    - All HW/Software Components
- **Triple Use Security Metadata**
    - Discovery, Marking, and Security Labeling/Credentialing
- **Trusted Binding Service**
- **Enterprise Sign-On and Session Management**
    - Global, Reduced, Single
    - WS-Aware and Legacy Applications
    - DoD Users and Applications and External Connections
- **Trusted Directories** - for Local, Enterprise, Global, Federated segments
- **Smart Audit**
    - Visibility of Security-Relevant Events at all Echelons
    - Dynamic ability to Correlate/Fuse
    - Preservation of Legal/National Security Evidence
    - Automatic Adjustment of Security Posture based on Audit-Enabled Indications and Warnings (I&W)

- **Multi-Echelon Digital Security Policy Management Console, Smart Push, Precedence**
- **Digital Security Policy Enforcement and Enterprise Visibility of Posture**
    - Service/Application "Firewalls"
    - Data "Firewalls" ex. guards, controlled interfaces, sanitizers, down-graders
    - Security Configuration Agents
- **Key Management Service**

## REVIEW OF CHAPTER GOALS

The goals of this chapter were to address:

1. How we can define information assurance and what the core information assurance principles are:
    - Define information assurance from DoDD 8500.1, and define the five information assurance principles.
2. What Defense-in-Depth means and how it is implemented as a DoD information assurance strategy:
    - Defense-in-depth provides a strategy based on people, technology and operations.
3. How the DIACAP provide certification and accreditation activities:
    - DIACAP involves a 5-step process that follows the information system life cycle.
4. What Common Criteria is and how security vendors are meant to comply with Common Criteria standards:
    - Define Common Criteria and discuss the different evaluation assurance levels that rate a vendor product based on security assurance requirements
5. The Net-Centric Information Assurance Vision and Goals:
    - Define the vision and state five goals for implementing information assurance:
6. How we can protect network information by using information assurance measures
    - Describe broad set of information assurance capabilities for net-centric operations.

## REFERENCES

Office of the Assistant Secretary of Defense for Networks and Information Integration, Department of Defense Chief Information Officer Directive 8320.02-P (2007, April). *Data Sharing in a Net-Centric Department of Defense.* National Security Agency (2003). *Defense in Depth - A practical strategy for achieving Information Assurance in today's highly networked environments.* Department of Defense Instruction 8510.01 (2007, November). *DoD Information Assurance Certification and Accreditation Process (DIACAP).* Department of Defense Instruction 8500.01E (2007, April). DoD Directive 8500.01E. *Information Assurance (IA).*

Office of the Assistant Secretary of Defense for Networks and Information Integration, Department of Defense Chief Information Officer (2007) *Department of Defense Information Sharing Strategy* (Version 1.0). Department of Defense Directive 8500.01E (2007, April). *Information Assurance*. Department of Defense Instruction 8500.2 (2003, February). *Information Assurance Implementation.*

Office of the Assistant Secretary of Defense for Networks and Information Integration, Department of Defense Chief Information Officer (2008, March). *The Department of Defense Interim Information Assurance Strategic Plan.*

Office of the Assistant Secretary of Defense for Networks and Information Integration, Department of Defense Chief Information Officer. (2008). *Department of Defense Information Management/ Information Technology Strategic Plan 2008-2009.* Bendel, M. (2006, March). *An Introduction to Department of Defense IA Certification and Accreditation Process (DIACAP).* Washington, DC: Lunarline Inc.

Turner, G. (2003, November). *Net Centric Enterprise Services Workshop on IA/Security*, OASD(NII).

# Chapter 10
# Adhering to Open Technology Standards

## CHAPTER CONTENT

As you explore through Chapter 10, it covers the following topics:

- The Need for Open Standards
- CASE STUDY: The Berkman Center
- Technology Standards Organizations
- Key Concepts for Technology Standards
- Defining a Technical Reference Model
- DoD Technical Standards Classification
- Representative System Technical Standards Profile

DOI: 10.4018/978-1-60566-854-3.ch010

## CHAPTER FOCUS

This chapter lets you understand why open technical standards are necessary. We then provide a case study of an organization who is working toward getting various entities and international officials together to coordinate the growing need for common standards. We bring up the topic of technology standards organizations that prepare and maintain technical standards for the greater society. Each technical standard has a set of stages through their life that includes development, draft, available, and sunset. We provide an example technology classification model that can be used as a reference for representing technical standards within an organization. After this we provide an understanding of the DoD Technical Reference Model and a representative system technical standards profile. The standards profile is comprehensive in nature, and provides a common set of technical standards that are in use at any organization.

## THE NEED FOR OPEN STANDARDS

In this current Internet age, it is important to discuss a vital concept that makes everything work and that is the use of "open standards". It is not necessary to sell the concept of open standards. Almost everyone agrees that they are necessary. The problem resides as to how they are implemented and how application and system vendors make money from products that they create using open standards.

Most participants in the information age recognize that open standards usually provide specifications that are owned and fostered by vendor-neutral organizations, and not one particular "proprietary" organization.

Open standards have been touted as the digital equivalent of a common gauge for railroad tracks. Open standards enable different information technology systems to communicate effectively when necessary. This recognizes that technology systems do spend most of their time consuming and processing data in their own specific manner.

Open standards and "open source" software is not the same thing. In the past decade, the IT industry has seen a large proliferation of open source software, which is really a development model for software in which code is freely shared and improved by a cooperative network of programmers.

Governments have also followed suit in quick adoption of this open source software since it allows software application development costs to be reduced dramatically. The spread of open source software allow the benefits of openly sharing information technology to reduce costs and make it easier for users themselves to innovate. Anyone is free to build open source software according to specifications without infringement of intellectual property rights, though there are limits to that, which is set by the software owner or maintenance vendor.

## CASE STUDY: THE BERKMAN CENTER

### Berkman Center For Internet & Society At Harvard University

A project begun by the Berkman Center for Internet and Society at the Harvard Law School (can be researched at http://cyber.law.harvard.edu/) put together officials from a number of countries around the world. These included officials from United States, China, India, Thailand, Denmark, Jordan,

Brazil and others. The group discussed technology standards and economic development, and focused on the use of technology standards to help foster growth in developing countries.

The group defined an open standard as technology that is not owned by a single company and is openly published. However, there is still a huge debate among public policy makers in the government, and non-governmental and non-profit organizations as to how far openness can go.

A report funded by the Berkman Center makes clear that governments around the world "should mandate technology choice, not software development models". The report also points out the following: ""Open source does not define an open information and communications technology ecosystem, but it can be an important transformative element,"."To date, open source has been the most disruptive element of the entire open agenda, provoking re-examination of information and communications technology ecosystems and policies."

Based on the Berkman Center mission on their web site, "The Berkman Center's mission is to explore and understand cyberspace; to study its development, dynamics, norms, and standards; and to assess the need or lack thereof for laws and sanctions."

*"We are a research center, premised on the observation that what we seek to learn is not already recorded. Our method is to build out into cyberspace, record data as we go, self-study, and share. Our mode is entrepreneurial nonprofit."*

Based on questions and answers provided within the Berkman Center website at http://cyber.law.harvard.edu/ we provide information here about the Berkman Center.

- When was the Berkman Center established?

In 1996, Harvard Law School professor Charles Nesson and Jonathan Zittrain established what was then called the "Center on Law and Technology" at Harvard Law School. The Center grew out of a seminar with Arthur Miller, David Marglin, and Tom Smuts in 1994 on cutting-edge Internet issues. The Center set out "to explore and understand cyberspace, its development, dynamics, norms, standards, and need or lack thereof for laws and sanctions." A gift of $5.4 million in 1997 from the Berkman family--Jack N. and Lillian R. Berkman, and their son Myles--underwrote Nesson's vision. Lawrence Lessig was awarded the Berkman professorship. In 1998, the Center changed its name to the Berkman Center for Internet & Society at Harvard Law School, and has steadily expanded ever since.

- Is the Berkman Center part of Harvard Law School?

Yes, it is. The Harvard Law faculty provides the mainstay of our research and Harvard Law students participate at all levels in our exploration of the relationship between Law (writ large) and cyberspace. We offer research support to faculty and teach a variety of cyberlaw courses--many in collaboration with other Harvard schools and MIT--integrating what we do with the larger mission of the University. We are not associated in any fashion, however, with admission to any of the Harvard Law School academic programs. To learn more about these programs, please visit the Harvard Law School Admissions Website. Please as well note that the Berkman Center does not award degrees.

- Does the Berkman Center provide legal assistance?

As a general rule, the Berkman Center does not (and cannot) provide legal assistance. However, Harvard Law School students occasionally contribute to cyber-related cases as clinical work, under the supervision of a Berkman Center-affiliated faculty member.

## TECHNOLOGY STANDARDS ORGANIZATIONS

At this time, there are a large number of technology standards organizations that provide the glue to ensuring commonality in our information field. Standards exist in any technical field, and most standards organizations are self-ruling bodies that have been chartered by corporate and public entities. A standards organization, standards body, or standards development organization (SDO) as they are widely-known is any entity whose primary activities are developing, coordinating, promulgating, revising, amending, reissuing, interpreting, or otherwise maintaining standards. These standards address the interests of a wide base of users amongst industries in different business market segments.

Technology standards can proliferate from a number of technology working groups who collaborate and share upcoming thoughts and proposals as to how information sharing can become more uniform. Most voluntary standards are provided for use by private organizations, mass consumers, industry regulators or governmental organizations. When the standards are then voted on and published, the standard may receive wide-spread acceptance. This then allows the technology standard to become dominant and can become the broader de facto standard for an industry. There are many examples of these phenomena, for example the Hayes modems specification, the Apple True Type font, and the Adobe PDF computer format.

Technology standards organizations include a wide number of examples such as the World Wide Web Consortium (W3C) and the Internet Engineering Task Force (IETF), who are the SDO for all the current and upcoming Internet standards, such as FTP, HTTP, and XML. There are international organizations such as the ISO (International Organization for Standardization) or the ITU (International Telecommunications Union) and IEC (International Electrotechnical Commission) who have been around for better part of the 20$^{th}$ Century. These standards organizations have established a large number of standards that have been adopted over time.

Standards organizations have been promoting open standards for a long time, and as the amount of information available to people grows by leaps and bounds, they recognize that their efforts become even more important over time. There are a number of areas that most standards organizations agree on:

- **Insist on market products and solutions that conform to Open Standards**

The use of open standards makes it easier to share information, overcomes all of the issues that arise with custom software interpretation, foster the need for greater interoperability based on time to integrate, costs to integrate, and risks to integrate

- **Only write Proprietary, or Custom Software when all other options have failed**

It is expensive to write custom software, a great deal of operational risk is introduced within applications, and the ability to decipher the software code is based on uniquely skilled personnel

- **Recognize that Technology Standards have a critical importance for the future**

As technology standards emerge, and they are adopted in the marketplace, their maturity increases their importance and their use within information technology; this becomes even more critical over time as people collaborate and interact outside their organizational and national boundaries

- **Standards can only play a big role if this potential is realized across organizations**

The use of technology standards normalizes the phenomena of data development, information dissemination, and information collaboration and sharing; however, this can only be done if organizations promote the need for open standards and not always resort to the need for developing proprietary solutions from a handful of vendors

## KEY CONCEPTS FOR TECHNICAL STANDARDS

Standards are essential elements for the functioning of information technology hardware, software, and networks. Technical standards promote an open method to communicate between different system interfaces. The objective of using a standard interface permits disparate devices and applications to communicate and work together.

Compliance to technical standards is a must when information technology has to enable the permanence of long term digital archiving and preservation of digital assets. Technical standards promote interoperability among different automated information systems. Use of standards also illustrates the use of "best practices" in terms of use of current technology processes.

As technology evolves, technical standards also have their own life cycle for standards usage. Since archiving of information is usually beyond 25 years, it is important to note that a major technical standard may still only be mature and open for use for example, for only 10 years.

Most technical standards undergo a specific set of stages throughout their life. The stages are as follows:

- **Development** – this status means that the standard is under development, and it is seeking input from various organizations.
- **Draft** – this status means that the standard is now in draft stage and is ready for initial circulation.
- **Available** - this means that the standard is placed for circulation and is mature for the industry to use.
- **Sunset** - this status means that the standard is ready for replacement and can be superseded by new standards that are up and coming.

The important point to recognize is that standards in usage by various programs within the industry can be in any one of the above stages. Since digital assets have to be stored and retained for the long

*Figure 1. Example of a Technology Classification Model*

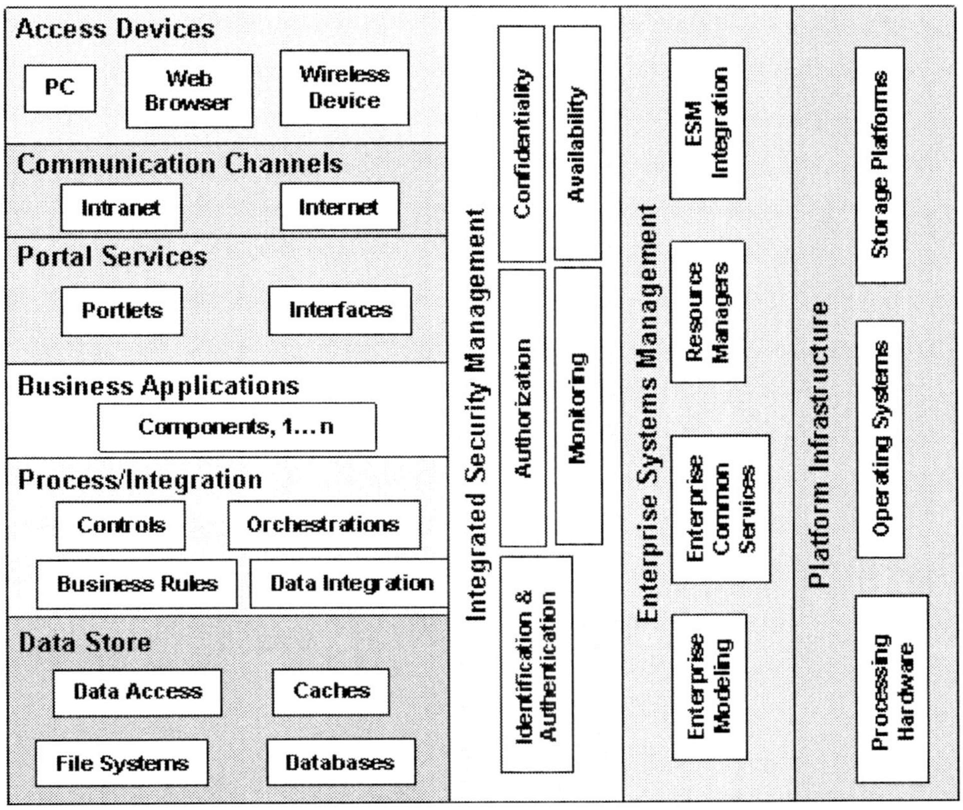


term, the retention periods are much longer than the stages for the technical standards. The only way to combat this difficulty is to count on the standards organizations that are the keepers for these standards. The responsibility of the standards organizations is to ensure that standards are kept up to date, and a continual change process occurs to combat technology evolution.

## DEFINING A TECHNICAL REFERENCE MODEL

The Federal Enterprise Architecture as prescribed by the OMB has prepared a FEA Technical Reference Model (TRM) (OMB FEA TRM v1.1, 2003) provides a foundation to describe the standards, specifications, and technologies supporting the secure delivery, exchange, and construction of business (or Service) components and e-Gov solutions. The FEA TRM unifies existing agency guidance by providing a foundation to advance the reuse of technology and component services from a government-wide perspective.

Adopting a consistent architectural framework against which information technology development efforts are reviewed and validated furthers the following enterprise goals:

- Allows ease of integration between applications, application services and enables inter-organization collaboration and sharing

*Table 1. DoD EA Technology Reference Model Classification*

| **Service Access and Delivery** |
|---|
| • **Access Channels** – Web Browser, Wireless or PDA Device, Collaboration/Communication, Other Electronic Channels<br>• **Delivery Channels** – Internet, Intranet, ExtraNet, Peer to Peer (P2P), Virtual Private Networks<br>• **Service Requirements** – Legislative/Compliance, Authentication/Single Sign-On, Hosting<br>• **Service Transport** – Network Services Transport |

- Increases the level of application interoperability within different levels of government and corporate firms
- Provides better responsiveness to changing business needs and rapidly evolving information technologies
- Allows faster deployment of new applications
- Provides efficient sharing and reuse of current technology assets
- Reduces the level of resources and costs required to develop, support and maintain the organization's applications
- Enables the consolidation of the information technology infrastructure to reduce costs, improve service levels, and increase operational flexibility across the enterprise

The following technology classification model provides an example of how technical standards can be aggregated within an engineering framework for development of hardware and software. This one takes advantage of Service Oriented Architecture principles to represent a target state that correlates system functions to different software, hardware and operations management components.

A layered pictorial illustration is presented in Figure 1 for this representative model.

Each of the components within the target Technology Classification Model is described in a vertical layer, based on what is first available to the end user:

- **Access Devices Layer** – specifies the end-user devices that is used to access the system
- **Communication Channels Layer** – specifies the communications methods and channels to be used for providing services to the access devices
- **Portal Services Layer** – specifies individual portlets, applications and interfaces that allow end-users to communicate and receive information
- **Business Application Layer** – specifies each of the various business applications that end-users can access to perform business functions within a system
- **Process/Integration Layer** – this layer includes the business process, workflow, orchestration, business rules, controls and data layer integration that is embedded within the ERA System.
- **Data Store Layer** – specifies the types of data stores that the ERA System will include such as caches, files systems, and databases and how the data will be accessed by the data stores.
- **Integrated Security Management** – this layer is vertically aligned to the classification stack that integrates within each layer, the concepts of identification and authentication, authorization, confidentiality, availability, and monitoring.
- **Enterprise Systems Management** – this is also a layer that affects the entire stack and provides enterprise modeling, common services, resource management, and ESM integration.

*Table 2. An Example system technical standards forecast*

| System Technical Standards Profile | | |
|---|---|---|
| **Technology Component** | **Technical Standards** | **Description** |
| **Web Browser** | HTML | HTML, the HyperText Markup Language is a standard for defining document type within the Web. |
| **Transfer Protocol** | HTTP | Web-based HyperText Transfer Protocol based messages for transfer of text graphics and images. |
| **Mail Protocol** | SMTP | Simple Mail Transfer Protocol (SMTP) for electronic mail transfer. |
| **Internet Protocol** | TCP/IP | Transmission Control Protocol/Internet Protocol to conduct Internet-based communication. |
| **LAN/MAN Protocols** | Ethernet; Gigabit Ethernet | IEEE 802.3 Ethernet standards and Gigabit Ethernet is defined by the IEEE 802.3z and 802.3ab standards. |
| **LAN/MAN Devices** | 10/100/1000 Base-T | Follows Ethernet and Gigabit Ethernet switch technology for all devices |
| **WAN Protocols** | Frame Relay; SONET | Frame-relay service from external third-party carriers.<br>SONET is a transport protocol for Fiber for offering OC-3 and OC-48 carrier lines. |
| **Application Server** | J2EE Server | JAVA 2 Enterprise Edition (J2EE) is an enterprise-level development environment for creating and deploying JAVA programming language and associated technologies. |
| **Portlet Specification** | JSR-168 | JAVA Specification Request (JSR) 168 establishes a standard API for creating portlets and enables interoperability between portlets and portals. |
| **Message Format Specification** | XML | XML is the current standard for message format and transmittal specification. |
| **Text Message Format** | RFC 822 | RFC 822; Standard for the format of Internet text messages provides the text message format. |
| **Message Invocation Specification** | SOAP | Simple Object Access Protocol (SOAP) used as an underlying protocol to send messages from one Application or Service to another within a Services Oriented Architecture. |
| **Service Description Specification** | WSDL | Web Service Description Language (WSDL) is an XML-based language for codifying services within a Services Oriented Architecture and offer web services for business functions. |
| **Locator Service Specification** | UDDI | Universal Description, Discovery and Integration (UDDI) is an XML-based registry that allows web services to be listed and seen by external business entities. Use of UDDI has to be implemented at a high-level so multiple systems can access a common registry. |
| **Remote Portal** | WSRP | Web Service for Remote Portals (WSRP) is a service specification published in UDDIs that allows remote portals and intermediary applications to integrate content and applications. |
| **Business process and work-flow specification** | BPEL | Business Process Execution Language (BPEL) is an XML-based language that is currently emerging to enable sharing amongst multiple organizations. |
| **Interactive Web Application** | WSIA | Web Services Language for Interactive Applications (WSIA) is to create an XML and web services centric framework for interactive web applications. |
| **Style Sheet Integration** | XSL | Extensible Style sheet Language (XSL) family allows XML transformation and presentation and acts as an integration adapter for web services. |
| **Access Control** | XACML | Use of eXtensible Access Control Markup Language (XACML) for implementing access control. |
| **XML Query** | XQuery | W3C XML Query provides flexible query facilities to extract data from real and virtual documents and collections both locally and on the Web. |
| **Integration Brokers** | JAVA RMI | Java Remote Method Invocation (RMI) system allows an object running in one Java Virtual Machine (VM) to invoke methods on an object running in another Java VM, and provides remote communications between programs. |

*continued on following page*

*Table 2. Continued*

| System Technical Standards Profile | | |
|---|---|---|
| **Technology Component** | **Technical Standards** | **Description** |
| **Business Rules Engines** | JSR-94 | JAVA Rules Engine API (JSR 94) defines a Java runtime API for rule engines by providing a simple API to access a rule engine from a Java Platform. |
| **Message Oriented Middleware** | JMS | The Java Message Service (JMS) API is a messaging standard that allows application components based on the Java 2 Platform, Enterprise Edition (J2EE) to create, send, receive, and read messages. |
| **File Transfer & Transport** | SecureFTP | SecureFTP is a file transfer client with great flexibility in configuration and transfer protocols. SecureFTP supports the secure SSH-1, SSH-2, and SSL/TLS protocols. |
| **Database Connector** | JDBC | Java DataBase Connectivity (JDBC) technology is an API (included in both J2SE and J2EE releases) that provides cross-DBMS connectivity to a wide range of SQL databases and access to other tabular data sources. |
| **Data Access** | JDO | Java Data Objects (JDO) API is a standard interface-based Java model abstraction of persistence, developed as Java Specification Request 12 JSR 12. |
| **XML Query Access** | XQJ | XQuery API for Java (XQJ) is a standard API that implements XQuery for Java and coordinates with JDBC. |
| **Digital Certificate Authority** | X.509 PKI | X.509 is an ITU-T standard for public key infrastructure (PKI). Digital Certificate Authorities are X.509-based that is offered by government agencies or commercial third-party providers. |
| **Authentication Servers** | IEEE 802.1-based | Authentication servers control authentication for enterprises via a centralized repository authentication information. IEEE 802.1-based authentication is used for Ethernet. |
| **Access Control Lists** | Windows-based ACLs, Solaris-based ACLs | The access control list (ACL) is a concept in computer security used to enforce privilege separation. |
| **Role & Policy Managers** | DoD and Services-based | Role and policy management is a working-level activity within DoD and Services to set the proper guidance. |
| **Perimeter Protection Technology** | Firewalls; IDS; Proxy Servers | Intrusion detection systems, firewalls, proxy servers and protection technologies are used to address this function. |
| **Network Virus Scans** | CERT | Computer Emergency Response Team (CERT) issues alerts and incident handling and avoidance guidelines. |
| **Desktop Virus Protection** | CERT | SUN and Windows-based desktop virus protection based on vulnerability alerts. |
| **Directory Services** | LDAP | Lightweight Directory Access Protocol (LDAP) is a "lightweight" version of a Directory Access Protocol (DAP) X.500 standard for directory services within a network. |
| **Secret Key Encryption** | FIPS 197 | FIPS 197 provides the Advanced Encryption Standard (AES) scheme for providing security encryption for ERA. |
| **Session Encryption** | SSL | Secure Sockets Layer (SSL)-based session encryption is the current widely used standard. |
| **Secure Media Deletion** | DoD 5220.22 | This is the DoD NISPOM standard for media deletion that involves expunging data. |
| **IPSec (Internet Protocol Security)** | IPSec standards | A full set of IP Security standards that are available for organizations to use. |
| **Checksum & Hash Algorithm** | SHA 256; SHA 512 | Checksum is a count of bits in a transmission unit that enables the receiver to verify if the number of bits received matches the number sent. Secure Hash Standard (SHA) - 512 and SHA 256 will be used. |
| **Intrusion Detection Systems** | | An Intrusion Detection System (or IDS) generally detects unwanted manipulations to systems. |

*continued on following page*

*Table 2. Continued*

| System Technical Standards Profile | | |
|---|---|---|
| **Technology Component** | **Technical Standards** | **Description** |
| **Vulnerability Assessment Tools** | CVE | Common Vulnerability and Exposures (CVE) - is a list of standardized names for vulnerabilities and other information security exposures. |
| **Event Correlation** | CoBIT | CoBIT is an IT governance framework and supporting toolset. We need to perform event correlation from the logs and alerts of multiple security technologies. |
| **Backup/Restore** | Standard Partition level Image | A Standard Partition Level Image is typically stored on tape, may contain multiple partitions and includes some overhead information in addition to the data being backed up. |
| **Remote Copy** | PPRC | Peer to Peer Remote Copy or PPRC is the protocol to mirror a DASD volume in one Control Unit (the primary) to a DASD volume in the other Control Unit (the secondary). |
| **Processor Clustering Technology** | RAC | Oracle Real Application Cluster (RAC) is a cluster database that works in Solaris environment with a shared cache architecture that overcomes the limitations of traditional shared-nothing and shared-disk approaches. |
| **System Management tools** | | System management tools allow the ERA enterprise to monitor the runtime condition of computer hardware platforms, operating systems, and monitor capacity. |
| **Archival Standard** | OAIS | Use of standard reference model for an Open Archival Information System (OAIS) provides archiving model for records preservation. |
| **Records Management Specification** | DoD 5015.2 | DoD 5015.2-STD, "Design Criteria Standard for Electronic Records Management Software Applications," provides implementing and procedural guidance on the management of records in the Department of Defense. |
| **Enterprise Modeling** | Zachman Enterprise Framework | Zachman Enterprise Architecture framework follows the NARA Enterprise Architecture and Levels 1 to 5 provide a model for system architecture. |
| **System Modeling** | UML | Unified Modeling Language (UML) is the industry-standard language for specifying, visualizing, constructing, and documenting the artifacts of software systems. |
| **Language Compilers** | JAVA | JAVA Language Specification for Solaris and JAVA Community Process (JCP) provides the current J2EE programming environment. |
| **Scripting Language** | JSP | Java Server Pages (JSP) is a Java technology that allows software developers to dynamically generate HTML, XML or other types of documents in response to a Web client request. |
| **Interactive Scripting** | AJAX | Asynchronous JavaScript And XML, or its acronym, AJAX, is a Web development technique for creating interactive web applications. |
| **Configuration Management & Version Control** | ANSI-EIA-649-1998 | ANSI/EIA-649-1998 National Consensus Standard for CM and MIL Handbook-61 provides the latest CM guidelines to be followed for the government enterprise. |
| **Requirements Management** | IEEE/ANSI 830-1993 | ANSI 830-1993 covers the recommended practice for software requirements specifications. |
| **Testing Software** | ANSI 1008-1987; ANSI 829-1983; ANSI 1012-1986 | ANSI has several standards related to software testing: 1008 deals with unit testing, 829 deals with verification and validation and 1012 is the standard for software verification and validation plans. |
| **Workstation** | Windows XP | Operating system for desktop workstations. |
| **Midrange Servers** | Solaris; Windows Server 2003; Linux ES4 | Operating system for ERA mid-range servers. |
| **High-End Servers** | Solaris | Operating system for high-end servers. |
| **Workstation DBMS** | SQL 1999; SQL/CLI 1995 | Desktop level standard for database operations. |

*continued on following page*

*Table 2. Continued*

| System Technical Standards Profile | | |
|---|---|---|
| **Technology Component** | **Technical Standards** | **Description** |
| **High-End Server DBMS** | SQL 1999; SQL 2003; SQL/CLI 1995 | Enterprise level standard for database operations. |
| **Workstation - Solaris** | Ultra SPARC, AMD x86 | Use of SUN Fire workstations. |
| **Workstation - Windows** | Pentium-based, AMD x86 | Use of the latest Pentium or x86-based standard for use for new purchases. |
| **Mid-range Server - Windows** | Pentium-based, AMD x86 | Use of latest Pentium or x86-based server. |
| **Mid-range Server - Solaris** | Ultra SPARC | Use of SUN Fire T2000 server using the UltraSPARC T1 processor. |
| **Servers (High End)** | Ultra SPARC | Use of SUN Fire 890 server with 8 CPUs. |
| **Server Input/Output** | PCI(e) | PCI Express, or PCI-E, is an implementation of the PCI computer bus, but bases it on a completely different and much faster serial physical-layer communications protocol. |
| **Cable Plant** | 100 TIA/EIA 568 | 100 TIA/EIA 568 is a standard for providing RJ-45 connectors to CAT5 cables. |
| **Fiber-optic cabling** | 100Base-FX | 100Base-FX over Fiber provides the standard for fiber cables necessary for network carriers and third-party subscriber lines. |
| **Storage Subsystems (SAN-based)** | Fibre Channel | Fibre Channel ANSI X3.230-1994 is a technology standard for transferring data between computer devices. |
| **Storage Subsystem (NAS-based)** | Gigabit Ethernet | NAS is a way of attaching storage to standard networks, and a NAS appliance is a file server that runs on Gigabit Ethernet and uses a file-sharing protocol to let clients access the storage. |
| **Redundant Disk** | RAID-5, RAID DP | Redundant Array of Independent Disk (RAID) Level 5 has a rotating parity array that allows all read/write activities to overlap; RAID Double Parity allows further protection. |

- **Platform Infrastructure** – this layer provides the infrastructure to the entire stack and includes server hardware, processors, operating systems, storage hardware and storage networks.

## DoD TECHNICAL STANDARDS CLASSIFICATION

Here we address DoD technical standards classification by providing the structure as to how the DoD Technical Reference Model (TRM) (DoD EA TRM v0.4, 2004) has been put together. The DoD TRM is organized into four core service areas, each with a supporting set of service categories, and each service category with supporting standards. Each Service Area aggregates and groups the standards, specifications, and technologies into lower-level functional areas. This parallels the TRM organization structure promoted by the FEA, and allows the DoD TRM to align closely with the FEA TRM.

Table 1 provides a categorized inventory as to how technical standards should be placed within the DoD TRM.

A full set of the latest technology standards can be found within DoD guidance documents. Table 1 in chapter 8 provides an understanding of how standards should be organized so that it fosters an environment of adopting open standards.

## REPRESENTATIVE SYSTEM TECHNICAL STANDARDS PROFILE

The following technical standards profile can be used as a representative set in any organization that promotes open standards. Table 2 was prepared to address the many technical standards that are aggregated at large organizations.

Here is a caveat, since hardware and software technologies have a fairly short time span, the following technical standards may become obsolete in the next few years. The following list provides a representative sample at an exact point in time. Use of these standards within the named categories is meant to allow greater interoperability between ongoing systems and promote the adherence to organizational needs and principles.

## REVIEW OF CHAPTER GOALS

The goals of this chapter were to address:

1. A need for open technical standards:
   - Evaluate how proprietary technical standards can transition to an open format
2. Some of the activities of the Berkman Center:
   - Provides a center on law and technology within Harvard Law School
3. What most technical standards organizations agree on, with regard to standards adoption:
   - Describe the roles technical standards play within organizations
4. That technical standards have their own life cycle:
   - This includes being in the development, draft, available, and sunset stages
5. An understanding of a technical reference model for a large organizations:
   - Review the technical services, and category of standards that are provided within the DoD technical reference model
6. A representative technical standards profile for an organization:
   - Review the common standards that are listed, their technology component from the classification model, and the name and description of the standard

## REFERENCES

Berkman Center for Internet and Society at the Harvard Law School. (2009). *Forum: Technology and the Internet*. Retrieved from http://cyber.law.harvard.edu/about

Department of Defense. (2004). *DoD Enterprise Architecture Technical Reference Model (TRM) version 0.4*. Retrieved from http://www.defenselink.mil/

Federal Enterprise Architecture Program Management Office (2003, August). *The Technical Reference Model (TRM)*, (Version 1.1).

Ghosh, S. (2006, July). *Electronic Records Archive (ERA) Technology Evolution Plan*. National Archives and Records Administration, Lockheed Martin, 2nd ERA Technology Evolution Council Meeting.

Ghosh, S., & Janczewski, M. (2005). *Standardization Concept Paper (SCP) VA/DoD Digital Imaging Initiative (DII)*. Joint Incentive Fund (JIF).

Krechmer, K. (2005). The Meaning of Open Standards. *International Journal of IT Standards and Standardization Research*, *4*(1).

Morris, E., Levine, L., Meyers, C., Place, P., & Plakosh, D., (2004, April). *System of Systems Interoperability (SOSI): Final Report*. Technical Report, CMU/SEI-2004-TR-004.

Simcoe, T. (2005). *Open Standards and Intellectual Property Rights*. Oxford, UK: Oxford University Press.

Sutor, B. (2006-2009) *Open Source vs. Open Standards.* Retrieved from http://sutor.com/newsite/drupal/osvsos

# Chapter 11
# Service-Oriented Architecture and Net-Centric Principles

## CHAPTER CONTENT

As you explore Chapter 11, it will cover the following topics:

- Service-Oriented Architecture for the Enterprise
- Key Service Oriented Architecture Concepts
- Benefits of SOA Adoption
- Service-Oriented Information Assurance
- Net-Centric Goals for Service-Oriented Architecture
- Net-Centric SOA Principles

## CHAPTER FOCUS

This chapter introduces Service-Oriented Architecture and puts it in the context of the net-centric future. We discuss key service-oriented architecture concepts as it applies to the information industry and provide

DOI: 10.4018/978-1-60566-854-3.ch011

a set of open technical standards that everyone needs to adhere. We then define a set of SOA terminology and provide a description of each of the elements that make up a SOA within a large enterprise. We discuss the benefits of SOA adoption, how it helps application configuration and interoperability, and what points the enterprise needs to consider. Then we provide a set of DoD stated net-centric goals that helps to move forward from the current view of the enterprise to the future view. We discuss the net-centric SOA principles and SOA governance activities as it relates to large organizations.

## SERVICE-ORIENTED ARCHITECTURE FOR THE ENTERPRISE

We have been using this buzzword throughout this book, and it is time to further define the term and provide context as to why this is such an important concept for achieving Net Centricity. Service-Oriented Architecture (SOA) is an information technology strategy that organizes the discrete functions contained in software applications into interoperable, standards-based services that can be combined and reused quickly to meet business user needs.

There is now a further concept of a service-oriented enterprise that poses to be the next-generation infrastructure state to allow the DoD and the military to achieve its design strategy for a net-centric future. A SOA environment helps military organizations to leverage their existing assets to manage the transition to network-centric warfare. For commercial organization, a SOA environment allows a direct exchange between service consumers and service producers. SOA serves as an overall framework that helps to provide the following functionality:

- Helps align information technology resources and personnel with organizational principles and practices by allowing software applications to inherit the goals and mission of the enterprise
- Allows organizations to gain the ability to create software solutions that deliver increased levels of agility, efficiency, and flexibility for constant modifications
- Allows the enterprise to align current technology with business needs and stakeholder concerns

Much of the power for SOA implementation begins with a basic yet powerful element, and that is the "Web Service". A web service is a self-contained package of code that provides a defined functionality – this means that this independent piece of software by itself does something useful.

According to the Organization for Advancement of Structured Information Standards (OASIS):

*"A Web Service is a software component that is described via WSDL and is capable of being accessed via standard network protocols such as, but not limited to, SOAP over HTTP."*

In most settings, a web service can be a single command to perform scheduled maintenance on a system or facility, a mechanism to prepare design changes to an engineering unit, or ability to provide a different assignment to resource personnel. The web service piece of code needs to work within a business process, and needs to be orchestrated properly to create a useful software application.

A service-driven enterprise is optimized around a particular set of services. This includes efficiently servicing customers, partners and employees, and accelerating the service response time of the business. In a large enterprise that has adopted a service-oriented architecture, it is expected that the automated process provides the business logic so that services are acted upon and combined to quickly achieve the desired result. Small pieces of code all codified separately as Web services can be aggregated together within the enterprise to provide industrial strength functionality. Since the DoD is such a complex organization, it helps the enterprise create objects custom to the end-user needs quickly to match the enterprise-level requirements. This may then lead to providing the needed services such as materiel management or force planning.

## KEY SERVICE ORIENTED ARCHITECTURE CONCEPTS

It is important to remember that SOA is a collection of services that communicates via a high-level abstraction layer. SOA exchanges occur using existing and emerging industry Web services interoperability standards. An organization taking an SOA approach would plan to provide or obtain a service or a business function over the Web using a prescribed set of Web services standards.

The following set of Web services standards form the foundation of SOA development:

- **HTTP** - Protocol for distributed, collaborative, hypermedia information systems
- **HTML** - A markup language for presenting information from the Web to end users
- **XML** - A common markup language for exchanging information
- **SOAP** – An XML-based specification for defining how Web services exchange messages
- **WSDL** – An XML-based taxonomy for defining the characteristics and functionality of a web service
- **UDDI** – Provides a mechanism for Discovery: a central repository which lists Web services that are available, akin to an address book

Figure 1 provides a quick look as to the interactions within a simple SOA environment that connects the Web Service Consumer with the Web Service Consumer with the Web Service Provider based on a common Service Registry.

1. The Service Provider registers and publishes the Web Service that it has created within the Service Registry
2. A Service Consumer who is the end user of the Web Service needs to perform a lookup request to find the particular service from the Service Registry
3. There is now a binding established between the Service Request performed by the Service Consumer and the Service Response from the Service Provider

Table 1 describes some of the commonly used SOA terminology that is associated with implementing a service-oriented environment.

A SOA environment usually includes the following set of elements as illustrated within Figure 2. The diagram in Figure 2 shows a simple hierarchy that includes a critical number of components.

These constitute the elements for an Enterprise SOA governance structure. Some of the definitions may be repetitive to the SOA terminology mentioned in the table above.

- **Application Front-end** – The graphical user interface that the end-user views on a computing device. There should be a mechanism to provide the service that the user is looking for, from their application front-end

*Figure 1. SOA Web Services in Action*

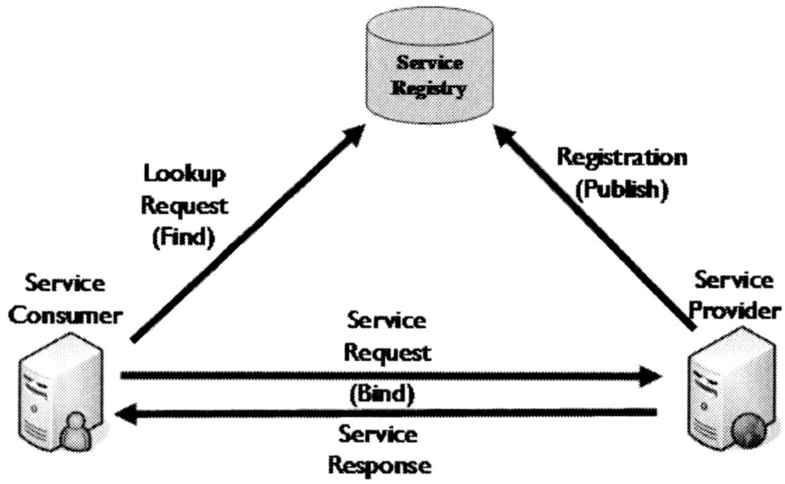


- **Service Repository** - A registry or repository where all of the assets of SOA are stored and catalogued
- **Service Bus** - The word "bus" is a reference to the physical bus that carries bits between devices in a computer. An enterprise service bus serves an analogous function at a higher level of abstraction. It allows routing and messaging capabilities in a SOA environment to transport the services
- **Service** – In this case, this is a Web Service or a group of Web services that in totality perform a service function for the customer. It contains within it:
  - **Service Contract** – Services adhere to a contract or communications agreement, that is defined collectively by one or more service description documents
  - **Service Implementation** – This signifies how the service is coded and implemented by the service provider to meet the needs of the service consumer

*Table 1. A Listing of SOA Terminology*

| | |
|---|---|
| **Service** | A function (business process) that is well-defined, self-contained and doesn't depend on the context or state of other services |
| **Service Provider** | Makes available a set of capabilities through a defined service contract for use by consumers |
| **Service Consumer** | Uses one or more published Provider services |
| **Service Contract** | The predefined agreement between a Consumer and Provider which provides interoperability between services |
| **Discovery** | Ability to look up the services available to Consumers |
| **Orchestration** | Coordination of services to execute a "process" involving a sequence of activities. Also known as Business Process Management (BPM) |
| **Loose coupling** | A design approach that decreases the impact of changes to services |
| **Web services** | The technical standards that enable communication between *services* in an SOA, which in turn combine to enable an end-to-end business process |

*Figure 2. Elements of a SOA-based Environment*

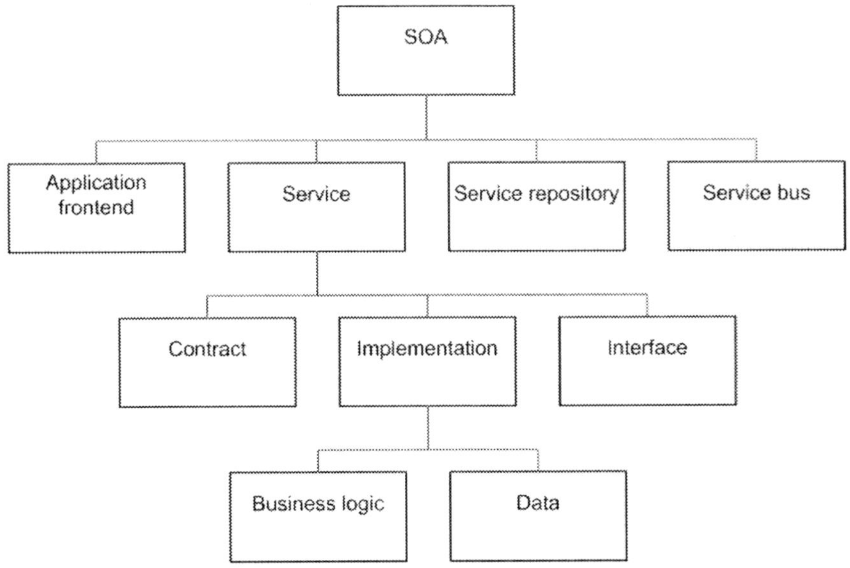


- **Service Interface** – A defined interface specification that hides the implementation of a language-specific service and allows a mechanism for the service to communicate data
- **Business Logic** - A set of policies that is meant to keep track of the "rules of engagement" for a distributed environment and meet a set of service levels expected in a SOA environment
- **Data** – The actual pieces of computer information passed within a service request that form the information exchange requirements for the enterprise.

## BENEFITS OF SOA ADOPTION

This section provides a broad understanding of why a service-oriented environment is the right answer to a number of the complex challenges that a number of organizations face including for example the government, financial, retail or transportation industry sector.

### Example Case for SOA Adoption

To realize why an enterprise should adopt a SOA-based architecture, it may be best to take a closer look at a specific environment. For the military, this might be to support field maintenance for a military organization. Within a military base, bringing a vehicle in for maintenance or repair can trigger a complex chain of events that may span many parallel organizations.

The steps or processes within this maintenance scenario, includes:

- Creation and filing of a work order,
- Accessing vehicle configuration documents,
- Ordering of materials

- Performing inventory checks,
- Documenting shipping orders
- Conducting status inquiries
- Researching parts replacement
- Updating the asset database

At this time, most organizations conduct these activities based on handwritten forms, faxes, or telephone calls. Based on the handoffs, this entire process may be time-consuming, costly and inefficient. Now if we analyzed the underlying applications and service-enabled it, then this process can be orchestrated as an automated process. This then becomes an enterprise SOA application that allows information to flow smoothly.

The results for this automated process would be the following:

- Participants will be able to view the status of the repair
- Process owners may be able to gain the ability to make continuous improvements to existing methods and procedures
- If maintenance directors decide to add processes to the repair scenario, such as repair scheduling, creating damage reports, or track invoices or payments – these applications can be linked in an automated manner quickly without costly programming services

## Enterprise Considerations for SOA Adoption

Large organizations need to consider a number of points before going ahead and moving toward the path of SOA adoption. The need to consider the following:

- **Projects & Applications** – At the beginning of the activity, the organization needs to identify existing projects and applications that are prime candidates for migrating to a SOA environment.
- **Information Architecture** – Organizations should develop an architectural framework for their enterprise that allows the assembly of components and services for the rapid and dynamic delivery of solutions. The information architecture needs to focus on software reuse and leverage existing assets and avoid the mentality of "stove-piped" applications or resources
- **Building Blocks** – SOA is both a architecture and a programming model since it is a way of building software. Engineering management within the organization needs to identify the key building blocks for this target enterprise that includes code, services, applications and components-that can be reused for a SOA implementation.
- **Business Strategy & Process** – The business stakeholder need to articulate how to align business and IT strategy so that business problems can be defined and solutions can be implemented in a coherent and repeatable way.
- **Costs & Benefits** – The compelling cost-benefit analysis case is that in a SOA-based environment, responsiveness can be improved while lowering development and maintenance costs.
- **Organization & Governance** – A SOA environment requires active organizations and governance which includes assigning roles and responsibilities for the new service-oriented IT support staff, and organizing an effective design for a reuse factory to atke advantage of existing software assets.

## Meeting Business Challenges

Tomorrow's business challenges will put great strains on personnel and the ensuing information technology environment. The pace is becoming more rapid where all of the business stakeholders including the customers, partners and employees all expect service on a 24 by 7 basis. This may require a new approach to conducting business activities within the enterprise.

The objectives of a SOA-based approach is to better align the information technology assets and resources with business goals so that the organization is better able to reuse assets, deliver value, and conduct activities faster so that they can more easily support ongoing requirements for change. To meet tomorrow's business challenges, the enterprise needs to:

- **Improve Customer Responsiveness:** Use the improved technology environment to gather and access information flowing through the business no matter where it is, and making it available to the people who need it to do their jobs. This allows you to be more responsive to customers and end users internally or externally
- **Improve Business Agility:** Rapidly adapt the business, and create internal and external touch points so that as the business changes you avoid having to begin from scratch with new applications and infrastructure as business requirements change.
- **Improve Operational Efficiency:** Make existing investments achieve higher productivity, which in turn, means improving the way business is conducted, and makes overall business processes more efficient.

From an information technology, hardware, software and infrastructure, a SOA-based environment hopes to provide the following fundamental benefits.

- **Reduce Complexity** – The adoption and adherence of technical standards enforces compatibility and avoids generic point-to-point integration, which thereby reduces complexity
- **Increase Reuse** – The creation, publishing and consuming of common services promotes the reuse of shared services that have been previously developed and deployed and allows more efficient application development and delivery
- **Improve Legacy Integration** – By analyzing, and interfacing legacy applications on a common service bus and offering a common service registry improves legacy migration and lowers the cost of maintenance and integration

## SOA-Based Information Security Principles

Within a SOA-based environment, all of us need an integrated information security environment. The essential functions of information security within a distributed SOA environment are to address the following:

- Provide an integrated information operation that protect and defend information and information systems
- Ensure that key information assurance principles such as availability, integrity, authentication, confidentiality, and non-repudiation are all properly verified within the SOA environment

- The integrated process needs to include protection, detection, and reaction, as well as the capability to incorporate information restoration.
- From the user perspective, provide a common operation picture containing unified assurance measurements to support cross-functional and aggregated effectiveness reports can be tailored to provide a higher degree of operational awareness

## NET-CENTRIC GOALS FOR SERVICE-ORIENTED ARCHITECTURE

Becoming net-centric and adopting service oriented architecture go together when trying to achieve the future state. According to the DoD Information Management/Information Technology (IM/IT) Strategic Plan for 2008-2009, the DoD goals moving forward can be stated as follows:

- **Net-Centric Transformation** - Accelerate DoD's net-centric transformation to facilitate effective and efficient warfighting, intelligence and business processes and other national security activities.
- **Information as a Strategic Asset** - Use information sharing to enable effective and agile decision making through visible, accessible, understandable and trusted data and services – when and where needed.
- **Interoperable Infrastructure** - Ensure robust and reliable world-wide connectivity and infrastructure within DoD and with external mission partners.

Based on these goals, the way to move forward is to clearly understand the definition of SOA in this context, the implication, and direct benefit to the organization.

- **Service Oriented Definition:** is an approach to building and managing a distributed computing infrastructure that configures the current information technology resources as services available and discoverable in a net-centric environment.
- **Service Oriented Implication:** Rather than dealing with isolated systems that must be integrated after the fact, Service Orientation provides operational users with understandable services they can call upon and compose into business processes as needed – building systems that can adapt as operations change.
- **Service Oriented Benefit:** Service Orientation provides agility and flexibility for users of technology, coupled with an abstraction layer that simplifies the complexity of today's heterogeneous IT environments.

The following diagram in Figure 3 provides an understanding of the DoD current view and DoD future view based on the use of SOA principles to transform the current state.

The current view shares in the problem of stove-piped or silo architectures where military organizations, for example, procure and develop this architecture to solve a specific problem. This is characterized by a limited focus and functionality and contains data that cannot be easily shared with other systems. Legacy systems that are bound by stove-piped architecture tends to become an assemblage of inter-related items that are so tightly coupled that the individual elements cannot be differentiated or upgraded and reused. This then increases the cost of technology refreshment since the entire legacy system needs to be thrown away and replaced by a new system.

*Figure 3. Net-Centric Goals based on Service Oriented Architecture*

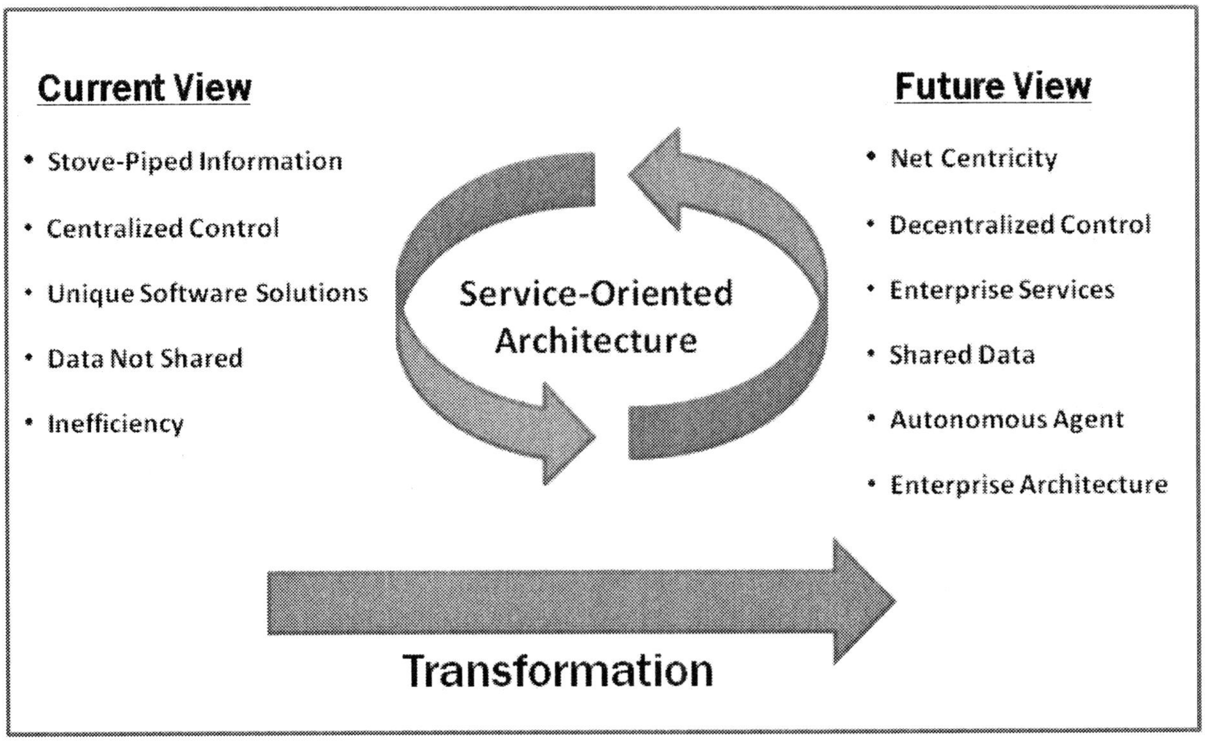


A SOA-based environment brings in a new paradigm for organizing and utilizing distributed capabilities that initially may be configured and controlled by multiple owners in different domains. SOA provides the means to organize solutions that promotes reuse, growth and interoperability.

## NET-CENTRIC SOA PRINCIPLES

Now we intend to combine the net-centric transformation and the ability to achieve a service-oriented architecture into a broad set of principles. So within this context, we can provide a full set of features that fit this new environment.

- **Loose Coupling** – Services maintain a relationship that minimizes dependencies requiring that they only retain awareness of each other.
- **Service Contract** – Services adhere to a communications agreement defined collectively by one or more service descriptions
- **Autonomy** – Services have control over the logic that they encapsulate
- **Abstraction** – Beyond that described in the service contract, services hide their logic from the outside world
- **Reusability** – Logic divided into services can be coordinated and assembled to form composite services

- **Statelessness** – Services minimize information retention specific to an activity, reinforcing the loose coupling principle
- **Discoverability** – Services are designed to be outwardly descriptive so that they can be found and accessed via discovery mechanisms

Based on the DoD Net-Centric Data Strategy that has been discussed in earlier chapters, we need to re-examine the data strategy from the SOA context. It is best to restate the data strategy goals in a manner so that we can all work with the data.

Here are a set of seven questions that is proper to ask:

1) **Who has what data available?**
   - This addresses the issue of data visibility allows us to coalesce information in a SOA environment
2) **Where is this data and in what format?**
   - This allows us to data accessibility across a large, distributed organization
3) **What does this data mean?**
   - This helps the data become more understandable
4) **Is this data trustworthy, accurate, and authoritative?**
   - This allows us to build on our information assurance strategy across the enterprise
5) **Can my application use the data?**
   - Each set of users have to answer this question and see how interoperable each of their applications are
6) **Is the data applicable and timely?**
   - This question asks how the future environment can be more responsive to user needs by providing information that is timely and applicable
7) **What and who governs the definition, lifecycle, and use of this data?**
   - This answers the issue of data becoming institutionalized within the overall governance structure of the enterprise

In answering these questions, we want to now provide a glimpse as to how a SOA-based environment can purport to address these challenges. The architecture framework of a SOA-based enterprise should prescribe certain ways in which systems and capabilities are built and how they should interact with one another.

Here are the most applicable mechanisms for a SOA-based enterprise to work within the context of the DoD or any major governmental or corporate organization:

- **Reliance on open standards to promote greater interoperability** – When building new system, the consensus to promote architecture that is based on open standards for improved interoperability and reduced reliance on proprietary platforms, programming languages, and products.
- **Leverage the reuse of legacy systems**— In most organizations, there is already a large number of institutional information systems that provide critical data to the users. It would be facetious to rip them out and replace them with a brand new system. These so called legacy systems instead need to be fully evaluated for their current purpose and should be wrapped within encapsulated services that expose and share the data and valuable functionality contained within them.

- **Promote loose coupling based on enterprise-level interface specifications**— The reason to adopt loose coupling is so that when a web service functions within its own context, these interactions do not need to necessarily be exposed throughout the entire enterprise. This allows changes to be made to local systems without directly affecting enterprise-level specifications. Within the DoD current view, systems have used point-to-point integration by defining non-proprietary interface specifications. This point-to-point communication does not lend itself well in a distributed environment, unless business rules are written specifically for this purpose. A SOA-based environment can bring in loose-coupling that ensures that the degree of dependency between services and their consumers is kept to a minimum. Internal implementation details do not need to be shared by all parties, and this can be part of the service contract set up between the consumer and the provider.
- **Use of object-oriented patterns to promote system agility**— A SOA environment uses specific design patterns that allow web services to conduct themselves based on the given architecture. Object oriented technology allows services to inherit properties from other services, so that they can all be composed together and orchestrated within an automated workflow. The reason this increases the agility of the system is because it reduces the time and effort to create new services and applications or adapt to real-time changes. This certainly helps in theater or battlefield, and allows the military organization to better cope with an unknown set of parameters.

## Net-Centric SOA Governance

Since the topic of promoting web services throughout an enterprise is complex, this new net-centric environment requires a deliberate governance activity. This governance needs to be clearly understood and delineated by the leadership within the organization.

A SOA governance activity requires how the management of the enterprise will consider SOA in their supervision, monitoring, control and direction of the effort. SOA governance activities can be set up from the beginning to address these two aspects:

- Establish an organizational structure that determines who is empowered to make decisions with regard to information management and information technology to fully implement the future SOA-based enterprise
- Set up the necessary mechanisms, standards and processes that would be used by the organization in a distributed enterprise to control and measure the way decisions are implemented

It should be recognized that this governance mechanism is a key activity to the success of the future enterprise. A number of challenges come to mind that will be on the plate of the personnel within the governance board. These challenges include:

- Walk that fine line in providing sufficient policy and process infrastructure without jeopardizing the less regulated environment needed for a SOA environment to succeed
- Promote within the organization the ability to increase agility in building and deploying systems, by seamlessly integrating a host of legacy and heterogeneous entities
- Ensure that policy and management for a fast-paced delivery is understood and institutionalized so that systems can be changed and deployed quickly to meet the future needs of our military and external organizations

Here are a set of principles to follow for a SOA governance strategy for a large enterprise, to march forward from the current state of the enterprise to its future state.

- Prepare an organizational framework that will understands the principles of SOA governance, and led management own this as an initiative that has clearly defined responsibilities and objectives.
- Follow good program management practices to set objectives, capture metrics and measure results
- Understand program management duties by identifying and actively managing risks that crop up as a result of the new SOA environment
- Align the SOA activities with the organization's mission and goals and its business transformation ideas for the future
- Analyze current gaps in existing projects so that you can define target areas where SOA can be implemented and maintained
- Develop a mechanism to capture improvement strategies that heightens the governance of high priority areas

## REVIEW OF CHAPTER GOALS

The goals of this chapter were:

1. To evaluate a service-oriented architecture view of the enterprise
   - Describe the essence of SOA and how it changes the manner of current information technology approaches
2. To remember key SOA concepts and technical standards
   - Analyze the set of open technology standards that make up a SOA environment
   - Evaluate the common terminology associated with SOA elements
3. To measure the benefits of SOA adoption for the enterprise
   - Review scenarios of how business processes can be changed and automated based on the use of automated services
   - Meet the ongoing business challenges customer relationship management and operational efficiency
4. To promote the core aspects of sharing knowledge and assuring connectivity
   - Address shared knowledge, collaboration, technical connectivity, distributed decision making, and network management principles
5. To identify net-centric goals for a service-oriented environment.
   - Understand the results of a service-oriented definition, implication and benefit
6. To review net-centric SOA principles and how it affects the organization's data strategy
   - Evaluate how data goals are better understood and the advent of a net-centric services strategy

## REFERENCES

Bass, T., Mabry, R., (2004, April). *Enterprise Architecture Reference Models: A Shared Vision for Service-Oriented Architectures*, (Version 0.81). IEEE Milcom 2004.

Croom, J. Lt. Gen. C. E. U.S. Air Force. (2006, July). *Service-Oriented Architectures in Net-Centric Operations*. Retrieved from http://www.stsc.hill.af.mil

Dirner, M., Yuan, E., & Blalock, J. (2007). *Realizing the Army Net-Centric Data Strategy (ANCDS) in a Service Oriented Architecture (SOA)*, Crosstalk, 07. Retrieved from http://www.stsc.hill.af.mil

Office of the Assistant Secretary of Defense for Networks and Information Integration, Department of Defense Chief Information Officer (2007). *Department of Defense Net Centric Services Strategy* (Version 1.0).

Office of the Assistant Secretary of Defense for Networks and Information Integration, Department of Defense Chief Information Officer. (2007, July). *Net-Centric Checklist* (Version 2.1.4).

Office of the Assistant Secretary of Defense for Networks and Information Integration, Department of Defense Chief Information Officer (2007). *Department of Defense Information Sharing Strategy* (Version 1.0).

Office of the Assistant Secretary of Defense for Networks and Information Integration, Department of Defense Chief Information Officer (2008). *Department of Defense Information Management/ Information Technology Strategic Plan 2008-2009.*

Solution Brief, S. A. P. (2008). *Enterprise Service Oriented Architecture – Design, Development, and Deployment*. Retrieved from www.sap.com/contactsap

Solution Brief, S. A. P. (2008). *Enterprise Service Oriented Architecture for Defense.* Retrieved from www.sap.com/contactsap

United States Army Enterprise Solutions Competency Center. (2007, September). *U.S. Army Service Oriented Architecture Life Cycle Management Model*, (Version 0.5b).

Valdes, R., (2008, June). *Portals: The Original SOA Framework*. Gartner Internet Platform and Web Services.

# Chapter 12
# Transition to IPv6-Based Networks

## CHAPTER CONTENT

As you explore Chapter 12, it will cover the following topics:

- The Need for a New IPv6 Protocol
- IPv6 Network Basics
- IPv6 Support for Net Centricity
- Federal Mandate for IPv6 Transition
- DoD IPv6 Standards Guidance
- Transition Strategies from IPv4 to IPv6
- Securing Data Transmission using IPSec
- A Review of Mobile IPv6 Networking

DOI: 10.4018/978-1-60566-854-3.ch012

## CHAPTER FOCUS

This chapter addresses how the transition to the IPv6 protocol is part of the upcoming network and communications upgrade for enabling a net-centric environment. We discuss limitations of current IPv4 protocol, features of the IPv6 protocol and discuss IPv6 network basics. We then discuss IPv6 support for net centricity, and the federal mandate for IPv6 transition. DoD has developed a set of IPv6 guidance that provides IPv6-capable definition, IPv6 base requirements and IPv6 product classes. We discuss transition strategies from IPv4 to IPv6 and long term strategy challenges such as security implications. We provide data can be secured using the IPSec protocol and we go ahead and provide a review of mobile IPv6 networking.

## THE NEED FOR A NEW IPV6 PROTOCOL

This chapter focuses on the crucial issue of the upcoming changes to network and communication links that is at the heart of Net-Centric transformation. We all know that the present state of the Internet needs to be fully upgraded to accommodate the ongoing communications growth rate. The need is there for a complete overhaul of the current Internet, based on the wide-scale adoption of the IPv6 Protocol.

The current network protocol for the Internet follows Internet Protocol (IP) Version 4, which has been around for the past 30 years. Even though IPv4 has proven to be robust, interoperable, and easily scalable, it still has faced a number of limitations due to the enormous increase in usage within the past three decades. There has been an exponential growth in Internet-enabled hardware devices, software operating systems, and an increasing demand in sending secure communications traffic that far exceeds the current capacities of the IPv4 protocol-based networks.

The following provides a set of limitations of the current IPv4 protocol that is in use in most countries throughout the world.

### Limitations of IPv4 Protocol

- **The Lack of Usable Address Space**

The IPv4 protocol only allows a 32-bit capacity for static IP addresses. It forces a large organization such as a federal agency to use a technology called Network Address Translation (NAT). This allows network management personnel to map multiple private addresses to a single IP address. Even though this technology promotes conservation of address space, it does not support network layer security standards, and causes problems when one organization is trying to connect to other external organizations.

As the Internet has proliferated across the world, and address spaces have been parsed out, there are still large U.S. organizations such as the government and private universities which hold large sequences of addresses. However, these organizations have kept these addresses for their own use, causing greater scarcity for commercial organizations and their need for public IPv4 address space.

- **Allocation of Address Prefixes**

Based on the current methods for IPv4 address prefixes and how they have been allocated, central Internet backbone routers are routinely required to maintain very large routing tables, of over 85,000 specified routes. The IPv4 routing is based on both flat and hierarchical routing, and the current lookup table affects the ability for messages to be sent from location to location, and that slows down the overall speed of transmission.

- **Complexity of Device Configuration**

Most IPv4 addressing parameters need to be manually controlled within computer devices. Or, they allow a stateful address configuration control such as DHCP, Dynamic Host Control Protocol. As the number of computer devices proliferates there is a greater need for automatic control of IP addresses that do not rely on difficult-to-administer protocols such as DHCP.

- **Data Security Issues**

The IPv4 protocol did not include within its original set of technical standards, a method of securing the message data in transit. Over time, because of the need for secure communication between organizations, protocols such as Secure FTP, and Secure Socket Layer (SSL) have been used by government and commercial organizations to encrypt message data in transit. Recently, the IPSec (Internet Protocol Security) has been accepted as an add-on to the IPv4 protocol message that provides security for IPv4 packets.

- **Quality of Service (QoS)**

Since the IPv4 protocol was originally created to prevent World War III, it deliberately did not address the issue of quality of service, since in a doomsday scenario, it is most important for the message to just get to its destination. So while standards for QoS do exist for IPv4, no identification of packet flow for QoS handling by routers is present within the IPv4 header. Instead, there is a real-time traffic support for messages that relies on the IPv4 Type of Service (ToS) field and the identification of the payload, typically using a UDP or TCP port. However, this IPv4 ToS field has its limitations since a TCP port or UDP port payload identification is not possible when the IPv4 packet payload is encrypted.

So for the past decade, a number of organizations such as the World Wide Web Consortium, and world governments such as the United States, Japan, China, India and West Germany, have been keen on supporting the new IPv6 protocol adopted by the standards organizations. The IPv6 protocol was intentionally designed to handle the large scope of our Local Area Networks (LANs) and Wide Area Networks (WANs). It is designed to minimize the impact on upper and lower layer protocols by standardizing the packet header formation, which then makes it easy to handle new data types and ensures fast network performance.

## Key Features of the IPv6 Protocol

The following provides a set of features and benefits of the next-generation IPv6 Internet protocol.

- **A Very Large Address Space**

The IPv6 protocol provides an enormously large addressing space and network prefixes. This can now allow the allocation of large IPv6 address blocks to Internet Service Providers, commercial organizations and government agencies of every country in the world. IPv6 addressing allows enterprise level organizations to also aggregate the prefixes of its users or customers into a single prefix and announce this one prefix to the IPv6 Internet.

- **A Simpler Addressing Format**

The IPv6 header format may be larger, but it is simpler that the IPv4 header. The IPv6 header removes the IPv4 fields of Header Length (IHL), Identification, Flags, Fragment Offset, Header Checksum, and Padding. This speeds the processing of the basic IPv6 header. IPv6 implements a multi-level address hierarchy that reduces the size of the router table needed to be maintained, and thereby provides more efficient and scalable routing. All fields of the IPv6 header are also 64-bit aligned, which takes advantage of the current 64-bit computer processors.

- **Support for Auto-Configuration and Device Plug-and-Play**

Based on the increased need for computer devices that require plug-and-play auto-configuration, the IPv6 protocol provides a built-in auto-configuration feature. This feature enables a large number of IP hosts to easily discover the network and obtain new and globally unique IPv6 addresses. Use of this feature can now allow the deployment of a large number of devices such as cell phones, wireless devices and home appliances.

- **Embedded Support for the IPSec Protocol**

Since in the case of the IPv4 protocol, the IPSec standards were an add-on, the new IPv6 protocol mandates the IPSec standards and embeds it within the IPv6 protocol suite. The IPSec standard allows security extension headers that make it easier to provide encryption, authentication, and Virtual Private Networks (VPNs). The IPv6 protocol allows end-to-end security to be implemented based on security services for access control, confidentiality, and data integrity with less impact to the network.

- **Better Support for Mobile IP and Mobile Devices**

The new IPv6 protocol standard provides significant support to the mobile networks and mobile devices that allows devices to move around without breaking existing connections. Auto-configuration of mobile devices allows devices to obtain Care-Of-Address and eliminates Foreign Agent. The binding process allows the Correspondent Node to communicate directly with the Mobile Node, avoiding triangulation routing necessary within IPv4 networks.

*Figure 1. Comparison between IPv4 and IPv6 Address Schema*

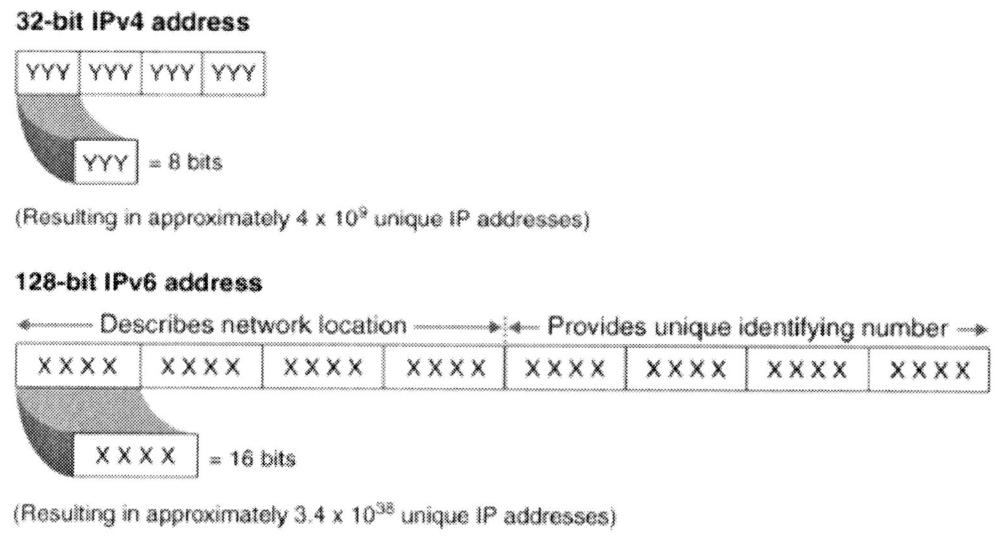


Source: GAO.

- **Elimination of Network Address Translation (NAT) activities**

As mentioned within the IPv4 limitations, NAT was used primarily to share and reuse the same address space among different network segments. Use of NAT within IPv4 networks temporarily solves the problem, but creates headaches when communicating with external networks. Since IPv6 address space is so large, there is no need for NAT translation, and eliminates the need for network address micro-management.

- **Support for a Number of Widely Deployed Routing Protocols**

*Figure 2. Schema for the IPv4 and IPv6 headers*

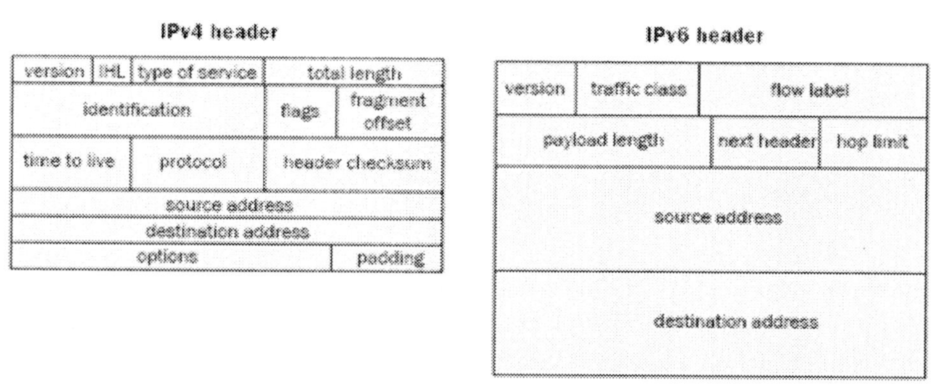

In the networking world in the past two decades, a number of protocols have been widely used for LAN and WAN connectivity. These include OSPFv3, IS-ISv6, RIPng and MBGP+ protocols. The IPv6 standards extend and maintain support for existing Internet Gateway Protocols (IGPs) and Exterior Gateway Protocols (EGPs).

- **An Increased Number of Multi-Cast Addresses**

IPv6 standards bring in multi-cast capability to deployed networks, provide for a number of multi-cast addresses and improve support for multi-cast. The IPv6 broadcast functions provide support for router discovery and router solicitation requests. The ability to multi-cast lowers network bandwidth requirements and improves network efficiency.

## IPV6 NETWORK BASICS

To gain a better understanding of the upcoming IPv6 standards, it is best to review a few topics on the IPv6 network format.

### IP Addressing Space

The IPv6 protocol standard dramatically increases the amount of IP address space available from the approximately 4.3 billion addresses in IPv4 to approximately $3.4 \times 10^{38}$. This is because IPv6 uses a 128-bit address scheme rather than a much smaller 32-bit address scheme used in IPv4. This allows many more possible addresses based on the actual bits in the address and the immense number of possible combinations of unique addresses. Figure 1 provides a comparison between the IPv4 and IPv6 address schema (GAO Report GAO-06-675, 2006).

Note that based on the linear concatenation of numbers, and multiple combinations, IPv6 can provide a gargantuan set of addresses.

### IP Header Information

The function of the IP header is to provide source and destination so that the message payload can successfully reach its destination. The IPv6 protocol simplified a number of complexities that existed within the current IPv4 header scheme. For example, the IPv6 header size is fixed. In the IPv4 case, the header size could vary complicating the readability of the message. Figure 2 provides the format for both the IPv4 and IPv6 header formats (Ixia IPv6 Paper, 2006).

The following defines each of the elements within the IPv6 header format.

- **Version:** This is a 4-bit IP version, in this case = 6
- **Traffic Class:** This is a 8-bit traffic class field, also in IPv4
- **Flow Label:** There is a 20-bit flow label that is used to identify traffic control on the Quality of Service
- **Payload Length:** This is the length of the IPv6 payload and is a 16-bit unsigned integer
- **Next Header:** This is a 8-bit selector that is used to identify the type of header that follows the IPv6 header

- **Hop Limit:** An 8-bit unsigned integer that decrements by 1 as nodes forward a packet.
- **Source Address:** Signifies the 128-bit address of the originator of the packet
- **Destination Address:** Signifies the 128-bit address of the intended recipient of the packet

### IPv6 Address Format

In using the new IPv6 protocol, network management personnel will have to get used to a new address format that will be embedded into all Internet-based hardware devices. The 128-bit IPv6 address is separated into eight 16-bit hexadecimal numbers divided by colons (:).

The preferred IPv6 address format is

XXXX : XXXX : XXXX : XXXX : XXXX : XXXX : XXXX : XXXX

For example: **3123:0000:1F1G:0000:0000:0100:11A0:ADFD**

#### Addressing Scheme for IPv6 Networking

In the case of IPv6, it is also important to become used to the fact that unlike an IPv4 node, an IPv6 node allows more than one type of IP address: Unicast, Anycast, and Multicast.

- **Unicast:** This is an address that is used to identify a single interface. This means that when a packet destined for a unicast address is delivered to the interface identified by that address.
- **Anycast:** In the IPv6 standard, the anycast address is designated as a global address that is assigned to a set of interfaces belonging to different nodes. A packet destined to an anycast address is routed to the nearest interface. The restrictions are that the Anycast address must not be used as a source address for the IPv6 packet, and it must not be assigned as the address of the IPv6 host.
- **Multicast:** In this case, a multicast address is assigned to a set of interfaces belonging to different nodes. This is no different from IPv4; however, a packet destined to a multicast address is routed to all interfaces identified by that address.

## IPV6 SUPPORT FOR NET CENTRICITY

As mentioned throughout this book, each chapter topic links itself to the global topic at hand of Net Centricity. Based on the Net-Centric Operational Warfare model, the DoD has been pursuing a technology upgrade to broadly address the Global Information Grid architecture. This architecture uses as its foundation, the new IPv6 protocol-based network. The IPv6 standards intend to greatly improve interoperability between all Internet-enabled hardware devices. These could be large entities such as weapons systems, tanks, aircraft, or floating vessels. Or they could be computing devices and embedded electronics that provide sensory capability to the military in the theater of operations.

DoD intends to use the increased address space provided by the IPv6 standard to greatly increase its reach within the warfighting and combat support operations through the Net-Centric information abilities. A visual example would be that the DoD can use the increased address space to render a three-dimensional map of the globe, based on IP addresses as coordinates. Each address would then be tracked and as part of the GIG architecture, the net-centric information will be stored, and relayed as necessary on a real-time basis.

The mobility of Internet-based devices is essential since the IPv6-based networks will allow enhanced mobility and auto-configuration to rapidly deploy networks across the globe. The IPv6 model will also allow a return to end-to-end communications based on information being kept secure at rest and secure in motion based on data encryption. The IPv6 communications security model will allow DoD to provide greater information assurance by providing more secure peer-to-peer communications.

The military is also trying to take advantage of the IPv6 Quality of Service standards to deploy messages at different priorities. Applications such as Voice over IP are currently proliferated in the commercial market, such as telephone applications for consumers to talk to one another. However, because of the lack of the quality of service, all messages are sent and received at equal priority, causing central routers to be flooded with incoming messages. The IPv6 Quality of Service standards will allow messages to follow a direct point-to-point communications within the confines of a distributed network topology.

A number of upcoming IPv6-based devices are being explored, that have potential within the Net-Centric environment, along with the commercial marketplace.

- **Hand-held Devices** – that take advantage of IPv6 mobility features to expedite the delivery of real-time data gathered from the field; this may include during field surveys and questionnaires, on-site investigations of industry and the work of revenue officers, security officers, auditors, and inspectors
- **Device Auto-Configuration** – the use of IPv6 auto-configuration features for operations planning and improve technology response time
- **End-to-End Security** - the use of the end-to-end security feature of IPv6 to build more secure retail and wholesale transactions, including securities and commemoratives
- **Network Management Costs** – lower the manpower costs of managing networks, and the communications infrastructure based on the improved network layer IPv6 standards

## FEDERAL MANDATE FOR IPV6 TRANSITION

Based on Net-Centric demands, the DoD has gone ahead and taken the lead within government agencies in trying to fully understand the deployment of an IPv6-based network across the globe. The IPv6 network initiative has been crucial for a military transformation, but most other Federal or State governments have not yet been as forward-thinking in their information infrastructure strategies.

Back in 2005, the U.S. Congress recognized that a majority of the 24 major Federal agencies other than the DoD had reported that they had not yet initiated key planning efforts for IPv6. This led to the recommendation that Presidential directives be initiated for this major technology initiative. The Director of the Presidential Office of Management & Budget (OMB) was instructed to take note of key steps in addressing IPv6 planning and security challenges. This includes the recognition of the following elements:

- IPv6-capable software and equipment need to exists in all Federal agency networks
- Federal agencies need to develop inventories of current network equipment and assess risks for upgrading the current base of technology

- Create a business case for IPv6 transition throughout government networks
- Establish government-wide policies and enforcement mechanisms
- Determine the overall cost for transition, including hardware and software
- Identify timelines and methods for transition

Based on these set of principles, the OMB issued a memorandum to all Federal Chief Information Officers through the Federal CIO Council, that a specific set of series of activities and associated deadlines need to be followed (OMB IPv6 Memo, 2005). This guidance was to ensure that Federal agencies are able to configure their infrastructure, specifically their "network backbone" by calendar half-year, June, 2008.

Each federal agency has had to assign officials to lead the transition activity from IPv4-based networks to IPv6 standards. The OMB memo required that an inventory is conducted to aggregate the set of routers, switches, and hardware firewalls. Agencies were also asked to develop an analysis of fiscal and operational impacts and risks of transitioning to a new IPv6-based network. Agency policies and enforcement mechanisms, training material, the ability to maintain and monitor government networks needed to be documented in an IPv6 Transition Plan.

This IPv6 transition plan document was required to be submitted to the OMB on a periodic basis. It is necessary to correlate this IPv6 Transition Plan with the existing government agency Current State and Target State Enterprise Architecture. It is expected that agency enterprise architecture planning is significantly affected by the transition of the network infrastructure. A vision of the future state of the Federal Enterprise needs to take account of the transition to the Next Generation Internet.

Based on the June 2008 OMB federal deadline, DoD Net-Centric mandates were made to be in convergence. DoD has been converting all digital military communications to use the new IPv6 Internet protocol as the standard for the Global Information Grid. It is also important to note that this IPv6 transition is not a U.S-based phenomenon. A large number of countries in Asia, Europe and Latin America are moving ahead with their own IPv6 initiatives. U.S. Federal government effort needs to be coordinated with Allied nations such as NATO countries so that network interoperability is addressed at a global level.

## DOD IPV6 STANDARDS GUIDANCE

The DoD has been working on IPv6 standards as part of government working groups, to address the definitive needs of the DoD Enterprise Architecture and Technical Standards View. Each DoD Component organization along with the Armed Force Service organizations have needed definitions for network and communications standards. The DoD IT Standards Registry (DISR) has published a DoD IPv6 Standards Profile, based on the guidance of the DoD IPv6 Transition Office (Green, D., 2005).

The idea is to create an "IPv6 Capable" definition for networking and communications equipment along with associated hardware and software. The IPv6-Capable definition brings up the concept of Technological Interoperability with regard to the latest hardware or integrated system.

## IPv6-Capable Definition

For a product to be labeled IPv6-Capable from the DoD perspective, the product must meet the following IPv6 requirement, along with support requirements for one or more of the following product categories. These are denoted as:

1) **Base Requirement:** All IPv6-enabled products must meet the IPv6 base technical standards for interoperability
2) **Product Classes:** Products are classified according to their system function on an IPv6 network. One or more of the following is applicable to the system's function:
   - **Host or Workstation:** end nodes and their user applications
   - **Network Appliance:** these could be hand-held devices, PDAs, phones, IP-based sensors, or home automation controllers
   - **Router:** intermediate nodes for packet routing
   - **L-3 Switch:** intermediate node that performs Layer-2 switching on IPv6 packet and has Layer-3 management and routing functions
   - **Information Assurance Devices:** security devices providing some aspect of DoD IA control measures
   - **Network Server:** provides some type of IPv6 network service such as FTP, e-mail, or HTTP web service
3) **IPv6 Protocol Functional Categories:** Network layer protocol standards are grouped by their protocol functional categories.
   - **Transition Mechanisms (TM):** All IPv6 nodes that interoperates with IPv4 must support one or more transition mechanisms for interoperability
   - **Connection Technology (CT):** Link layer connection technology definitions are necessary for all nodes (for example: Ethernet, ATM, PPP)
   - **IPv6 Security Profile (IPsec):** Protocols to achieve basic IP Security (IPsec) for IPv6
   - **Quality of Service (QoS):** IPv6 protocols related to QoS signaling and services
   - **Mobility (MOB):** Requirements to achieve IPv6 Mobility (MIPv6) and Network Mobility (NEMO) for IPv6
   - **Wireless Systems (Wireless):** IPv6 support for wireless systems including header compression and MANET routing
   - **Network Management (NM):** Protocols for network management in IPv6 networks

In addition to meeting all of the above set of needs, a product must support the IPv6 version of any IPv6 protocol functional categories, for DoD Net-centric purposes as required for its function within the DoD Global Information Grid (GIG).

## IPv6 Base Requirements

The DoD IPv6 Standards Profile has worked with technical standards organizations along with IPv6 testing organizations to decide on a complete set of technical standards. Based on the working group definitions, all IPv6 nodes must support the base requirements stated below.

*Figure 3. Types of IPv6 Capable Devices*

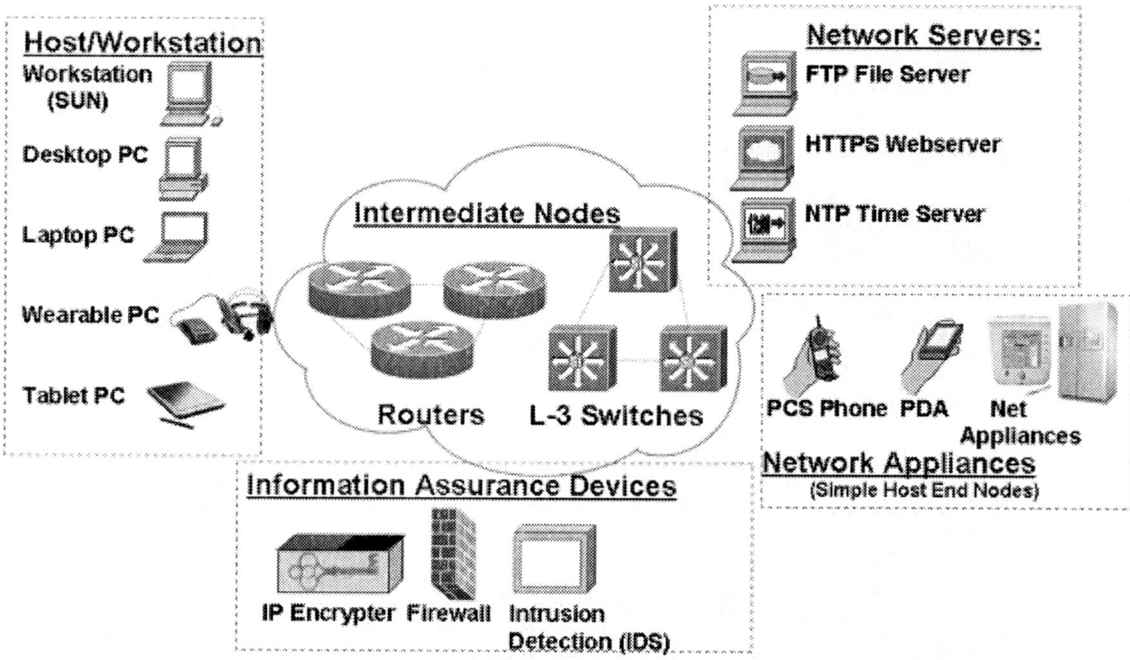


These now form the basis for DoD's Net-Centric certification as they form the core of interoperability requirements for IPv6 nodes connecting to the DoD GIG and Internet.

- RFC 1981, Path MTU Discovery for IPv6
- RFC 2460, Internet Protocol v6 (IPv6) Specification
- RFC 2461, Neighbor Discovery for IPv6
- RFC 2462, IPv6 Stateless Address Auto-configuration:
  - All nodes MUST support RFC 2462 auto-configuration of link local addresses
  - All end-nodes (non-routers) MUST support full RFC 2462 auto-configuration
  - Hosts/workstations and routers MUST support manual configuration and MUST support disabling auto-configuration
    - RFC 4443, Internet Control Message Protocol (ICMPv6)
    - RFC 4301, Security Architecture for the Internet Protocol: All end nodes and intermediate nodes requiring security MUST have an upgrade path to full IPsec by 2007. RFC 4301 specifies many other security protocols – See the full IPv6 IPsec Security profile in this document
- RFC 4291, IPv6 Addressing Architecture
- RFC 4007, Scoped Address Architecture (All IPv6 addressing plans MUST use this standard definition for scoped addressing architectures)
- RFC 4193, Unique Local IPv6 Unicast Addresses (Replaces the site-local address with a new type of address that is private to an organization, yet unique across all of the sites of the organization)

## IPv6 Product Classes

For the purposes of the IPv6 standard, the hardware, software and networking equipment can be broken down into the following product classes and ensuing example:

- Host/Workstation: End Node (ex. Personal computer)
- Network Appliance: End Node (ex. IP-based Phone)
- Router: Intermediate Node (ex. Network router)
- Layer 3 Switch: Intermediate Node (ex. Communications switch)
- Network Server: Simple Server and Advanced Server (ex. DHCPv6 Server, or HTTP web server)
- Information Assurance Device: Security Device (ex. Firewall, Intrusion Detection System)

The following diagram in Figure 3 provides a great illustration of the types of IPv6 capable devices.

## TRANSITION STRATEGIES FROM IPV4 TO IPV6

Transitioning from the current IPv4-based network to a next generation Internet is a considerable challenge, since there has been such an explosion of growth in the past thirty years. The Internet, by definition, is a global activity that crosses the network path of multiple countries, in a single message transfer.

A number of challenges exist that may slow the global adoption of the IPv6 standard. Switching to the new standard means that software applications that now use Internet addresses may need to be changed. Every Web browser, every computer, every email application, and every Web server must be upgraded to handle the 128-bit address for IPv6. The routers that operate the Internet backbone now implement IPv4 via computer hardware, and cannot route IPv6 over the same hardware. By adding software to route IPv6 packets, the routers will operate more slowly, which may cripple the Internet. Alternatively, upgrading and replacing the hardware for millions of Internet routers would be very costly.

Based on the complexity of the network transformation, it is expected that IPv6 deployment will be a coordinated, long term process. Most government and corporate organizations have to prepare for a long term transition strategy for their network communications standards. This means that even though IPv6 provides many benefits over the legacy IPv4 technology, all agree that a successful IPv6 deployment plan requires it to co-exist with IPv4 for an extended period of time. The rest of this section provides a number of planned strategies for transition of the IPv4 to IPv6 network infrastructure.

### Dual Stack Backbone

In a dual stack backbone deployment, all routers need to maintain both the IPv4 and IPv6 protocol stacks within the same network. This means that software applications riding on top of the network layer must choose between using IPv4 and IPv6. The operating system would select the correct IP address based on the type of IP traffic and network configuration.

At this time, having a dual stack backbone routed network is the preferred approach for transition, since it allows the flexibility of IPv4 and IPv6 applications. However, there are a number of limitations.

- All WAN routers must be upgraded to the IPv6 standard
- Routers require a dual addressing scheme, IPv6 and IPv4 address spaces
- Dual management of IPv4 and IPv6 traffic
- Sufficient memory for both the IPv4 and IPv6 routing tables

## IPv6 over IPv4 Tunneling

An IPv6 over IPv4 tunneling activity encapsulates the IPv6 traffic within the IPv4 packets. This merged packet can then be sent within an existing IPv4 network. The IPv6 protocol standard can be kept within an island, and end systems and routers can communicate throughout the IPv4 infrastructure. A number of different tunneling techniques are available:

- **Manually Configured Tunnels** – As defined by RFC 2893, both end points of the tunnel need to be configured with appropriate IPv4 and IPv6 addresses. The edge router, which should be dual stacked can then forward the tunneled traffic based on the configuration.
- **Generic Routing Encapsulation (GRE) Tunnels** – This type of message transport allows the IPv6 packet to be transported over the IPv4 network, by encapsulating the packet properly and configuring it for send and acknowledgement.
- **6over4 tunnels** – As defined by RFC 2893, these tunnel activities automatically set up tunnels based on a IPv4-compatible IPv6 address. This means that the left-most 96 bits are all zeroes, followed by information in the remaining embedded 32-bit IPv4 address.
- **Intra-Site Automatic Tunnel Addressing Protocol (ISATAP) Tunnels** – This type of tunneling uses an IPv4 address embedded in an IPv6 address to automatically identify the setup and end point of the tunnel. This type of tunneling is prevalent within large campus WANs, such as a large government site or a large corporate campus.

## Proxying and Translation

This type of transition strategy recognizes the fact that when an IPv6 server host needs to access an IPv4-only service that some form of translation is necessary. In the case of an IPv4 Web Server, the most widely supported form of translation is a Web Proxy, or the use of a dual-stack application layer proxy server.

Within the technical standards community, techniques for application-agnostic translation of an IPv6 packet within an IPv4 server have been proposed at the lower layers. However, a number of these translation techniques are one time affairs, and cannot be scaled to be used within government or large corporate networks.

## Long Term Strategy Challenges

Based on the transition strategies that were offered above, it is clear that government and corporate organizations have to apply multiple IPv6 and IPv4 network solutions. This means that a dual stacked environment will also require use of tunneling techniques, and proxy translation techniques.

The upcoming long term challenges faced by industry and government agencies include the following thoughts:

- **Use of a Continued IPv4/IP6 Dual Environment** – Government and corporate organizations need to recognize that dual IPv4 and IPv6 environments need to be maintained for an extended period of time. This means that most network personnel would have to maintain two network protocols, which adds significant complexity to the network environment. Maintaining system interoperability within this dual network will also be difficult, along with finding skilled personnel to manage the effort.
- **Use of the Multi-Homing Feature within IPv6 will be a Challenge** – Multi-homing occurs when a host is assigned an IP address from multiple Internet Service Providers (ISPs), who can provide the host with multiple addresses. This multi-homing feature gives organizations greater reliability, because if one ISP stops working, the organization can automatically switch to another provider. Currently, the method for implementing a multi-homing in an IPv4 environment creates routing issues in an IPv6 environment. This is an issue that the technical standards organizations are still debating about, and should come to an amicable solution.
- **Implementing new IPv6 Standards** – It is important to recognize that IPv6 standards still have not reached to a mature level. The current IETF organization is just working on RFC standards that are being categorized from Emerging to Mature. This means that these emerging standards may be interpreted by different network vendors differently. This can then lead to interoperability problems unless large government organizations such as DoD, or large corporate vendors provide guidance and strategy to mediate a common approach. This issue is also paramount when hardware and network devices are undergoing Conformance Testing. It is important to recognize that standards are an evolving process and each manufacturer and software developer follow a prolonged software development life cycle to bring their products up to the recognized standard. IPv6 standards have to be addressed in this manner within the context of the product development life cycle.

Internet Protocol version 6 (IPv6) contains numerous features that make it attractive from a security standpoint. It is reliable and easy to set up, with automatic configuration. The large number of address schemes renders it highly resistant to malicious scans and prevents automated, scanning and self-propagating worms and hybrid threats.

## Security Implications for Transition to IPv6

IPv6 is not a complete solution for security issues since few security problems derive solely from the IP layer in the network model. For example, IPv6 does not protect against mis-configured servers, poorly designed applications, or poorly protected sites. In addition, IPv6 and IPv6 transitional mechanisms introduce new, not yet widely understood tools and techniques that intruders can use to secure unauthorized activity from detection.

These IPv6-derived efforts are often successful even against existing IPv4 networks. Since many network administrators have yet to take advantage of IPv6, they may be unaware of IPv6 traffic that has tunneled into their networks. Attackers are already using this potential oversight to establish safe havens for attack. Fortunately, existing protection technology is equipped for IPv6, making protection across this emerging standard both practical and straightforward. The key is to enable administrators to protect against attacks, intrusions and backdoors that take specific advantage of the protocol.

The risks introduced primarily by IPv6 transition mechanisms can be mitigated and controlled using existing applications and techniques. Unconstrained 6to4 tunneling methods should be blocked by system administrators and security personnel based on a need-only basis. 6to4 is a transitional mechanism intended for individual independent nodes to connect IPv6 over the greater Internet. Any large network should provide orchestrated routing for IPv6 where IPv6 is supported, so there should be no need for unconstrained tunneling techniques.

Fortunately, most good firewalls will block IP protocol, the 6to4, and ISATAP tunnel protocol, unless it has been explicitly enabled. Thus, most firewalls by default block GRE tunnels and 6to4 tunnels. Security administrators need to verify and test this setting for tunneling mechanisms.

If IPv6 is supported, the common tunneling techniques should be provided by the infrastructure system (routers and gateways) whether or not 6to4 auto-tunneling is supported. Delivery of IPv6 traffic to individual nodes and workstations should be done via native IPv6. If tunneling techniques are supported, it should be supported intentionally from a limited number of well-defined gateways. This procedure provides direct access support from individual 6to4 nodes on the external networks while controlling the traffic flows on the inner, protected network.

## SECURING DATA TRANSMISSION USING IPSEC

This section provides an overview of Internet Protocol Security (IPSec) topics since it is the most widely used secure network protocol for use within an IPv6 environment. The IPv6 standards mandate use of the IPSec protocol, and require its use in all IPv6 implementations in the future.

### Header Protocols

IPSec consists of two packet encapsulation protocols—the Authentication Header (AH) that allows authentication of the sender; and the Encapsulating Security Payload (ESP) that supports both authentication of the sender and encryption of data. IPSec provides VPN capabilities at Layer 3 of the Open System Interconnection (OSI) model, whereas PPTP and L2TP operate at Layer 2. AH primarily provides packet integrity services, while ESP provides packet confidentiality services. In addition, IPSec supports two encryption modes—transport and tunnel. Transport mode encrypts the data portion (payload) of each packet, but does not encrypt the header. Tunnel mode encrypts both the header and the payload, making this mode more secure. In either mode, the receiving side of an IPSec compliant device decrypts each packet.

### Choosing Between IPSec Modes

IPSec operates in two modes: transport mode and tunnel mode. IPSec transport mode is used for host-to-host connections, and IPSec tunnel mode is used for network-to-network or host-to-network connections.

### Using IPSec Transport Mode

IPSec transport mode is fully routable, as long as the connection does not cross a network address translation (NAT) interface, which would invalidate the host network. Used this way, IPSec must be supported on both hosts, and each host must support the same authentication protocols and have compatible IPSec filters configured and assigned. IPSec transport mode is used to secure traffic from clients to hosts for connections where sensitive data is passed.

## Using IPSec Tunnel Mode

IPSec tunnel mode is used for network-to-network connections (IPSec tunnels between routers) or host-to-network connections (IPSec tunnels between a host and a router). Used this way, IPSec must be supported on both endpoints, and each endpoint must support the same authentication protocols and have compatible IPSec filters configured and assigned. IPSec tunnel mode is commonly used for site-to-site connections that cross public networks, such as the Internet.

## Selecting an IPSec Authentication Method

During the initial construction of the IPSec session—also known as the Internet Key Exchange, or IKE—each host or endpoint authenticates the other host or endpoint. When configuring IPSec, need to ensure that each host or endpoint supports the same authentication methods. IPSec supports three authentication methods:

- Kerberos
- X.509 certificates
- Pre-shared Key

## Authenticating with Kerberos

Kerberos is used for the IPSec mutual authentication, by default. For Kerberos to be used as the authentication protocol, either hosts or endpoints must receive Kerberos tickets from the same Directory Service. Kerberos should be chosen for IPSec authentication only when either the hosts or endpoints are within the same organization. Kerberos is an excellent authentication method for IPSec because it requires no additional configuration or network infrastructure. Based on IPSec exemptions, Kerberos is exempt from IPSec filtering.

## Authenticating with X.509 Certificates

X.509 certificates can be used for IPSec mutual authentication of hosts or endpoints. Certificates allow you to create IPSec secured sessions with hosts or endpoints outside your Active Directory forests, such as business partners in extranet scenarios. You also must use certificates when using IPSec to secure VPN connections made by using Layer Two Tunneling Protocol (L2TP). To use certificates, the hosts must be able to validate that the other's certificate is valid.

## Authenticating with Pre-Shared Key

It is best to use a pre-shared key that is a simple, case-sensitive text string, to authenticate hosts or endpoints. Pre-shared key authentication should be used only when testing or troubleshooting IPSec connectivity because the pre-shared key is not stored in a secure fashion by hosts or endpoints.

*Figure 4. Components of Mobile IPv6*

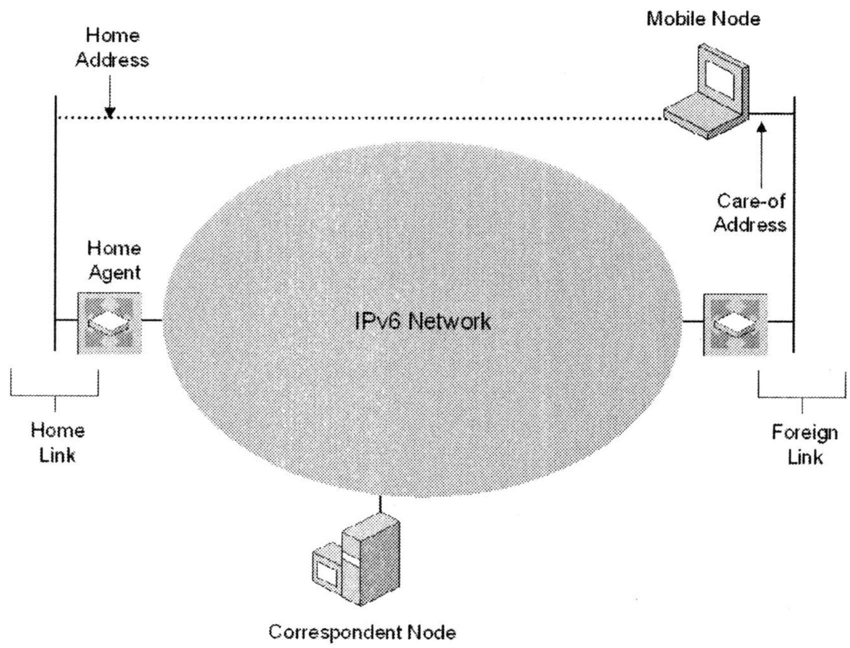


## Creating IPSec Policies

IPSec is a policy-driven technology. IPSec policies are dynamic, meaning that you do not have to stop and start the IPSec service or restart the computer when assigning or un-assigning IPSec policies.

**Client (Respond Only):** A computer configured with the Client policy will use IPSec if the host it is communicating with requests using IPSec and supports Kerberos authentication.

**Server (Request Security):** A computer configured with the Server policy will always attempt to negotiate IPSec but will permit unsecured communication with hosts that do not support IPSec. The Server policy permits unsecured ICMP traffic.

**Secure Server (Require Security):** A computer configured with the Secure Server policy will request that IPSec be used for all inbound and outbound connections. The computer will accept unencrypted packets but will always respond by using IPSec secured packets. The Secure Server policy permits unsecured ICMP traffic. In addition to the built-in policies, you can create custom IPSec policies. When creating your own IPSec policies, you must configure rules that include the following settings:

- IP Filter List
- Tunnel Settings
- Filter Actions
- Authentication Methods
- Connection Types

There are IPSec rules that determine the type of network traffic IPSec will initiate between the computer and the host. Since a computer can have any number of IPSec filters, you should ensure that only one rule is created for each type of traffic. If multiple filters apply to a given type of traffic, the most specific filter will be processed first.

## IP Filter List

The IP filter list defines the types of network traffic for each specific IPSec rule. You must define the following details for each entry in the filter list:

1. **Source address:** Can be a specific IP address, a specific IP subnet address, or any address.
2. **Destination address:** Can be a specific IP address, a specific IP subnet address, or any address.
3. **Protocol:** The protocol ID or transport protocol used by the protocol. For example, Point-to-Point Tunneling Protocol (PPTP) uses Generic Routing Encapsulation (GRE) packets. GRE packets are identified by their protocol ID, which is protocol ID 47. Telnet, on the other hand, uses TCP as its transport protocol, so an IPSec filter for Telnet would only define the protocol type as TCP.
4. **Source port:** If the protocol were to use TCP or UDP, the source port could be defined for the protected connection. The source port is set to a specific port or to a random port, depending on the protocol being defined. Most protocols use a random port for the source port.
5. **Destination port:** If the protocol uses TCP or UDP, the protocol uses a specific port at the server to accept transmissions. For example, Telnet configures the server to listen for connections on TCP port 23.

When configuring IP filter lists for transport mode connections, you should always choose to have the IPSec rule mirrored to secure the return communication defined in the rule. For tunnel mode connections, you must manually specify both the inbound and outbound filter list. The tunnel setting determines whether IPSec operates in transport or tunnel mode. If you want the filter to operate in tunnel mode, you must specify the IP address of the endpoint of the tunnel.

**Filter Actions**: For each filter rule, you must choose a filter action. The filter action defines how the traffic defined in the IP filter will be handled by the filter rule. The three filter actions are listed here and shown in chapter 9 figure 5.

**Permit:** Allows packets to be transmitted without IPSec protection. For example, Simple Network Management Protocol (SNMP) includes support for devices that might not be IPSec aware. Enabling IPSec for SNMP would cause a loss of network management capabilities for these devices. In a highly secure network, you could create an IPSec filter for SNMP and set the IPSec action to Permit to allow SNMP packets to be transmitted without IPSec protection.

**Block:** Discards packets. If the associated IPSec filter is matched, all packets with the block action defined are discarded.

**Negotiate Security:** Allows an administrator to define the desired encryption and integrity algorithms to secure data transmissions if an IPSec filter is matched.

## A REVIEW OF MOBILE IPV6 NETWORKING

The IPv6 standards also include a working set of mobile network specifications as organized by IETF and corporate vendor organizations. One reason for reviewing these standards is because mobile device technology is still in flux, and the future may bring a number of changes to the network landscape.

Mobile IPv6 technology allows an IPv6 node to be mobile, that is, to arbitrarily change its location on an IPv6 network, and still maintain existing connections. When an IPv6 node changes its location, it might also change its link. When an IPv6 node changes its link, its IPv6 address might also change in order to maintain connectivity. This chain of events needs to be tracked for the mobile routing to be completed.

Mobile IPv6 standards provide mechanisms to allow for the change in addresses when moving to a different link, such as for IPv6 stateful and stateless address auto-configuration. It is possible that when the address changes, the existing connections of the mobile node that is using the address assigned from the previously connected link cannot be maintained and are ungracefully terminated.

The key benefit to using Mobile IPv6 standards is that if the mobile node changes locations and addresses, the existing connections through which the mobile node is communicating are still maintained. To accomplish this, connections to mobile nodes are made with a specific address that is always assigned to the mobile node, and through which the mobile node is always reachable. Mobile IPv6 provides Transport layer connection survivability when a node moves from one link to another by performing address maintenance for mobile nodes at the Internet layer.

## Mobile IPv6 Components

Figure 4 shows the components of a Mobile IPv6 network.

Mobile IPv6 as shown in the diagram can be described as follows:

- **Home link** – This the link that is assigned to the home subnet prefix from which the mobile node obtains its home address. The home agent resides on the home link.
- **Home Address** - The mobile node is always reachable through this home address regardless of its location on an IPv6 network. If the mobile node is away from home (not attached to the home link), packets addressed to the mobile node's home address are intercepted by the home agent and tunneled to the mobile node's current location on an IPv6 network. Because the mobile node is always assigned the home address, it is always logically connected to the home link.
- **Home Agent** – This is a router on the home link that maintains registrations of mobile nodes that are away from home and the different addresses that they are currently using. If the mobile node is away from home, it registers its current address with the home agent, which tunnels data sent to the mobile node's home address to the mobile node's current address on an IPv6 network and forwards tunneled data sent by the mobile node. The home agent does not have to be the router that connects the home link to an IPv6 network. The home agent can also be a node on the home link that does not perform any forwarding when the mobile node is at home.
- **Mobile node** – This is an IPv6 node that can change links, and therefore addresses, and maintain reach-ability using its home address. A mobile node has awareness of its home address and the global address for the link to which it is attached (known as the Care-Of Address), and indicates its home address/care-of address mapping to the home agent and Mobile IPv6-capable nodes with which it is communicating.
- **Foreign Link** - A link that is not the mobile node's home link.
- **Care-of-Address** - An address used by a mobile node while it is attached to a foreign link. For stateless address configuration, the care-of address is a combination of the foreign subnet prefix

and an interface ID determined by the mobile node. A mobile node can be assigned multiple care-of addresses; however, only one care-of address is registered as the primary care-of address with the mobile node's home agent. The association of a home address with a care-of address for a mobile node is known as a *binding*. Correspondent nodes and home agents keep information on bindings in a binding cache.

- **Correspondent Node** - An IPv6 node that communicates with a mobile node. A correspondent node does not have to be Mobile IPv6-capable. If the correspondent node is Mobile IPv6-capable, it can also be a mobile node that is away from home.

## REVIEW OF CHAPTER GOALS

The goals of this chapter were to address:

1. The need for the entire global Internet network to be overhauled based on the new IPv6 network:
   - Provide limitations of the current IPv6 protocol and discuss how crucial features can be improved.
2. The key features of the new IPv6 protocol:
   - Discuss a host of new features of this IP version.
3. An understanding of IPv6 network basics:
   - Discuss addressing space, header information and address format.
4. TheIPv6 network to meet net-centric mandate and federal agency mandates:
   - Evaluate network management, mobile devices and upgrade of infrastructure backbone.
5. The guidance needed for meeting DoD IPv6 network mandates:
   - Define what it means to be IPv6 capable, IPv6 base requirements, and IPv6 product classes.
6. The transition strategies to IPv6 migration and how we can implement IPv6-based security:
   - Use of different migration techniques, dual stack and tunneling, address long term challenges, figure out how to secure data transmission using IPSec.
7. The requirements for IPv6-based mobile networking system:
   - Identify mobile components and discuss mobile internet standards.

## REFERENCES


3COM. (2004). *3COM IPv6 Technology Brief.* Retrieved from www.3com.com

Corporation, I. B. M. (2003). *IBM Vision for IPv6 – eBusiness*. Retrieved from http://www.software.ibm.com/ipv6

Digi, (2006, July). *A Primer on IPv6 White Paper*. Retrieved from www.digi.com

Executive Office of the President, Office of Management and Budget. (2005). *Transition Planning for Internet Protocol (IPv6)*, Memorandum for the Chief Information Officers.

Fritsche, W., (2000). *Mobile IPv6 – Mobility Support for the Next Generation Internet*, IPv6 Forum.

GAO Report to the Chairman, Government Reform, House of Representatives, GAO-06-675 (2006, June). *Internet Protocol Version 6*.

Ghosh, S., (2005, May). *Net Centric IPv6 Strategy for Theater Medical Information Program (TMIP)*, OASD(HA) Theater Medical Information Program.

Green, D., (2005, December). *DoD IPv6 Standard Profiles for IPv6 Capable Products*. Princeton, NJ: SRI International.

Ixia, (2006, November). *Internet Protocol version 6: Conformance and Performance Testing*, Ixia White Paper. Retrieved from www.ixia.com

Microsoft Corp. (2006). *Microsoft Windows Server 2003 - Introduction to IPv6*. Retrieved from www.microsoft.com

RFC 1752 (1995). The Recommendation for the IP Next Generation Protocol. W3C Consortium. Retrieved from www.w3c.org

# Chapter 13
# Storage Strategy for the Distributed Enterprise

## CHAPTER CONTENT

As you explore Chapter 13, it will cover the following topics:

- The Need for Greater Data Storage
- Determining Storage Requirements
- Storage Life Cycle Management
- Storage Architecture Review
- Enterprise Storage Architecture Design
- Enterprise Storage Management Activities

## CHAPTER FOCUS

We now switch to another crucial topic within a net-centric environment, and that is to address the digital storage needs within a global distributed environment. We address the need for greater storage; provide

DOI: 10.4018/978-1-60566-854-3.ch013

information as to how to determine storage requirements. We introduce the concept of storage life cycle management and discuss information management stages. A review of storage architecture is then provided which defines DAS, SAN and NAS. We go on to provide enterprise storage architecture design guidelines, and discuss the choices of centralized storage vs. distributed storage, vs. hybrid storage. We provide storage questions to consider for the enterprise and the net-centric challenges that lie ahead. We also discuss the people and processes and the dependency on network and communications.

## THE NEED FOR GREATER DATA STORAGE

Within the digital information enterprise of today and tomorrow, we have identified one challenge that is inherent to the growth of this industry. This challenge is to be able to store and manage increasing volumes of data. This is true for all of our government and corporate environments, including government contractors and commercial vendors. An enterprise storage strategy is a critical piece that needs to be addressed as part of the net-centric vision of the future for our military and civilian market.

Storage needs for the volume of data keeps increasing exponentially as we march forward toward our future information society. Industry experts report that storage needs have been increasing between 30 percent and 50 percent annually for the past two decades. This trend does not seem to be abating and probably will only soar as the size of multimedia data and file types increase by leaps and bounds. However, it is true that storage costs for digital storage and archiving is also decreasing dramatically, as hardware resources are able to contain more massive amounts of data. This is probably the reason why it has not occurred to us to do anything about this storage challenge – and so the problem seems to be resolving by itself.

We note that within the industry, no one is really caring about actual figures in storage volume, which may be in gigabytes, terabytes, or petabytes. What they are realizing is that the real challenge isn't really in storing the massive amounts of data. The real problem lies in managing that data, information, and knowledge across the enterprise. The entire information technology industry is working with business customers to figure out difficult issues such as how do we deal with storage management within business facilities, how do business customers have access to the data they have, how can they secure their data, and how can they backup their data for the future.

We all know that in the case of digital files, duplicate data is a huge issue since everyone wants to save their information multiple times. Individual users may keep multiple copies of a particular file in various stages of development. These files are also being shared with other staff members via e-mail, which in turn also automatically saves each version of the file. Unbeknownst to anyone, there can be dozens of copies of the exact same file being backed up every night for users at different locations. Now if we multiply this process across all of us across the globe, you are able to imagine how big this problem becomes.

From a large organization perspective, storing the internal data and information generated by the firm, is a mission critical activity that requires 24x7 availability, zero data loss, and rapid recovery times. With these types of requirements, organizations are spending more on storage, even though storage capacity is doubling each year. The rapid growth in the demand for data storage has many organizations re-evaluating the design and management of their information technology infrastructures. Specifically their enterprise storage needs, hardware capacity, and their ability to access their stored information across a far flung network.

## DETERMINING STORAGE REQUIREMENTS

Determining the storage needs of a large organization is very difficult activity. Certain industry professionals may actually say that it is a futile activity, since as soon as you come up with engineering estimates for enterprise storage, the technology industry makes these numbers obsolete, since the computing devices may have changed, the operating systems are different and the size of the data files are thereby much larger.

Hence, it is important to take an overall perspective to understand storage demands. Parameters such as availability, capacity, and performance need to be considered when determining storage requirements and evaluating storage architecture. To understand these demands, it is essential to collect accurate metrics and the criticality for data storage within the organization.

Estimation is an important aspect of the storage requirements and design process, and the quality of the metrics directly affects the ability to make meaningful estimates. In terms of physical metrics, one question to answer is how much capacity is required by the operating system. This is fairly simple to collect. However, what is more difficult is to evaluate the capacity requirements for various software applications being used within the organization. Personnel within the organization should be collecting physical metrics and logical metrics, indicating the actual volume of data and the external interfaces that affect the data storage requirements.

The volume of data used by different organizations can vary considerably. These days, the amount of data in messaging systems can vary because of elements such as different archiving and e-mail attachment policies that exist within the organizations. Software applications in use have varying degrees of storage requirements based on the entire software development life cycle. Physical metrics can be collected based on numerous tools that interrogate servers to gather physical metrics. Logical metrics deal with specifically with storage capacity requirements. Principal logical metrics cover availability, scalability, manageability, and performance. Each of these can directly affect the amount of storage capacity as well as throughput requirements.

A full set of storage requirements needs to examine the current information technology infrastructure that includes the number of computer systems and facilities. Inter-site storage and communications need to be integrated, but the organization's storage priorities may still be local and system-specific. Overall storage requirements need to consider a number of set factors for enterprise storage design and maintenance.

### Overall Storage Requirements for Large Organizations

Here are a set of key issues to evaluate when reviewing storage requirements:

- Data availability to address ongoing user needs
- Criticality of the data that is being stored
- Available budget for storage infrastructure
- Existing data volume and estimates for the future
- Enterprise network and communication capabilities
- Legacy storage systems
- Technical diversity of business systems
- Technical skills and resources

## Overall Storage Design Goals

In gathering storage requirements for future systems, the overall storage design goals should be to:

- Lower the total cost of ownership (TCO)
- Provide higher availability of data and information
- Ensure data is being protected and stored for example in encrypted format to ensure adequate information security guidelines
- Allow data storage to be incrementally scalable as more information is captured, processed and archived
- Provide a simplified management process to manage storage devices, stored applications, and user interactions for access and retrieval
- Ensure that when new storage systems are added later on, that data interoperability standards are preserved
- Avoid prescribing a single storage solution since this may likely conflict with future storage purchases and integration

## STORAGE LIFE CYCLE MANAGEMENT

Storage lifecycle management is a policy-based approach for managing and storing the flow of an information system's data throughout its lifecycle. The life cycle is characterized as follows:

1) The time of creation of the digital data placed in initial storage,
2) Retrieval based on users needs on an ongoing basis, and
3) A span of time until the data becomes obsolete and is deleted from the organization

Over the years, the storage industry has come up with a number of titles for this concept including Hierarchical Storage Management (HSM), Data Lifecycle Management (DLM) and Information Lifecycle Management (ILM). Each of these implementations is very similar in nature and has been adopted by the major storage vendors.

One such policy-based storage approach is ILM, which is currently popular in today's storage marketplace. ILM provides a comprehensive approach to managing the flow of a system's data and associated metadata from creation and initial storage, to the time when it becomes obsolete. ILM involves all aspects of dealing with data, starting with user practices. Storage products using ILM automate the processes that are involved and organize data into separate tiers.

According to policies specified within the storage infrastructure, ILM products automatically migrate data from one tier to another. As a rule, newer data, and data that must be accessed more frequently, is stored on faster, but more expensive storage media. Data that is less frequently accessed is placed on cheaper, slower media. The ILM approach however, recognizes that the importance of any data does not rely solely on its age or how often it is accessed. Users can specify data values based on how important the data is to the organization, how critical it is for survival, and how often it can be updated. Based on these different policies, the data can decline in value at different rates or can retain its data value throughout its life span. Within the storage management software, a path management application then makes it possible to retrieve any data stored by keeping track of where everything is in the storage cycle.

Storage life cycle management activities have become increasingly important in today's organizations, since businesses face compliance issues with regard to government legislation, such as HIPAA and the Sarbanes-Oxley Act. Business users are being asked to reproduce information from their past storage repository, for example, as part of a legal deposition. Within the DoD and military community, storage life cycle management is a critical activity to organize the vast amounts of data in multiple military facilities.

## Information Management Stages

Based on storage lifecycle management practices, the various information management stages can be categorized into a number of definitions. These include the concepts of data classification, data policy, data management, and tiered infrastructure.

- **Data Classification** - Incoming data is classified merely by user observation. An individual determines the class of the data, and then data is further classified by its value and relevancy. The value of the data changes regularly based on its age. The relevancy may change for example, due to multiple owners of the data is assigned.
- **Data Policy** - The changing value of data is taken into consideration and a set of policies is applied. Policy decisions may involve time that data remains in a particular location, the flow of the information, and the number of copies that should be stored.
- **Data Management** – The data management segment incorporate technologies that enable the actual placement, migration, replication, backup, archiving, and deletion of data based upon the policies defined by the data policy segment.
- **Data classification** - Usually, enterprise data falls into three broad categories:
  - Structured data that is placed for example in databases
  - Semi-structured that may be within an e-mail application
  - Unstructured data that is within desktop applications, audio or video files.

Within these broad categories, there is also data criticality. Within the enterprise, certain applications can be considered as follows:

- Mission critical, or gold-level applications
- Business critical or silver-level applications
- Operation critical, or bronze-level applications
  - **Tiered infrastructure** - A tiered storage infrastructure is how data is placed within storage devices. These may include a choice of high-end disk arrays, modular arrays, Serial ATA (SATA) drives, tape libraries, content-addressable storage (CAS) systems, archiving platforms, and Write-Once, Read-Many (WORM) tapes. Since storage devices may be one of the largest components of the overall infrastructure, and its effective utilization is critical to reducing total cost of ownership.

These information management stages allow enterprises to store their data in storage tiers. Based on cost estimates, business users may decide to keep the data based on its relative importance, reserving

*Figure 1. Different Storage Technology Configurations*

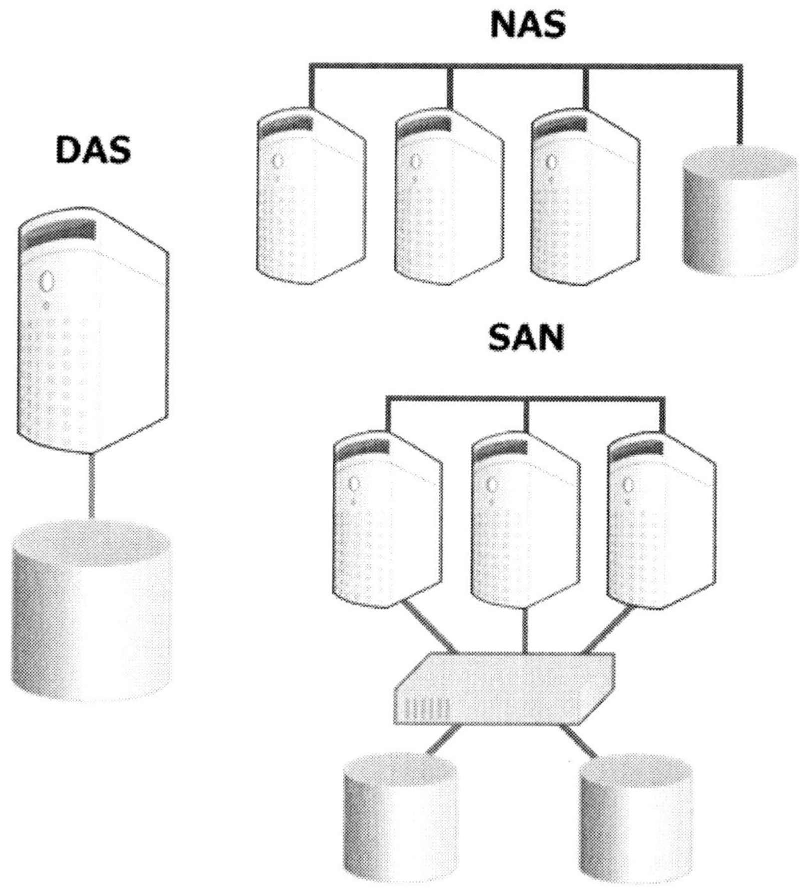


high-end arrays only for the most important data. Structuring the data into stages data also ensures that the data is organized by importance and usage and makes it better for preparing a disaster recovery plan.

## STORAGE ARCHITECTURE REVIEW

Storage architecture is fundamental to providing businesses the data they need on an ongoing basis. It is vital that the architecture enables organizations to quickly create secure, integrated business solutions that harness the data stored within hardware devices in the shortest timeframe and with minimum disruption of the existing business requirements. Storage architecture should also address the issue of data loss, since this can have an impact on the operations of the organization.

Decisions taken at the architectural level have a far-reaching impact on the overall enterprise. Many organizations today protect their storage assets as a vital business asset. This means that how data is stored and retrieved affects the design of systems and the services that they provide.

## Types of Storage Technologies

As part of the storage architecture review, we provide a description of the different types of storage technologies. These storage technologies are used to then put together an enterprise-level storage requirement. Figure 1 shows an icon-based description of each of the different storage technology architecture.

### Direct-Attached Storage (DAS)

DAS is digital storage that is directly connected to a computer server by a connectivity media such as a fiber or copper wire link. Some examples of DAS include the local disk drives that are often accessed through either an Integrated Device Electronics (IDE) or SCSI interfaces at the back of a computer. DAS may also include RAID (redundant array of independent disks) device controllers. The main characteristic of a DAS configuration is that it provides fast data access to the directly attached server. This however means that storage is accessible only from that single server.

### Network-Attached Storage (NAS)

A NAS device is a server that runs an operating system specifically designed for handling and storing data file services. The main characteristic of network-attached storage is that the storage is accessible directly on the local area network (LAN) through LAN protocols such as TCP/IP. NAS is a type of storage that is engineered to provide a flexible and scalable solution to the file-sharing needs of an organization. The downside of accessing storage using network protocols is that the speed of data access to the computing device is dependent on the responsiveness of the network infrastructure. This in turn, may affect end-user performance as compared DAS, where it is dependent on local bus speeds of the attached computer.

### Storage Area Networks (SANs)

A storage area network (SAN) is the next level of network storage management that provides a separate storage network specifically set up to provide access to high performance and highly available storage subsystems. The SAN is made up of specific devices called host bus adapters (HBAs) in the host servers, disk storage subsystems, tape libraries and SAN switches that help route storage traffic. The method of data transfer for a SAN switch is similar to those used by LAN network switches. All of these devices are interconnected by fiber or copper wire link.

The main characteristic of a SAN is that they are flexible and scalable based on the fact that storage subsystems are generally available to multiple hosts at the same time. The specialized nature of SAN HBAs and SAN switches are that they provide a performance benefit over NAS. But this also means that a SAN may also be more costly than a NAS and has more overhead that can break down. Although DAS data transfer rates are still faster, the performance gap between DAS and SAN technologies is consistently shrinking. The advantage of multiple servers being able to use the storage solution in a SAN means that it is more suited for more enterprise level needs.

## ENTERPRISE STORAGE ARCHITECTURE DESIGN

When it comes to a large enterprise, it is important to prepare for the next generation storage system, by performing storage architecture design. This design should follow a structured approach to ensure that the correct solution is adopted by the organization.

In most cases, you can categorize storage architectures into three basic types:

- Centralized Storage
- Distributed Storage
- Hybrid Storage

We can characterize these basic types of storage as logical design choices 1, 2, and 3.

### Logical Design Choice 1: Centralized Storage

The centralized storage design model defines a single storage solution that is best fit for the enterprise, and is powerful enough to meet everyone's needs.

- **Advantages for Centralized Storage**
  - The centralized storage model provides a lower total cost of ownership (TCO) because all of the storage needs can be provided by a single solution
  - A single storage model can also be made highly secure where the input and output can be access controlled and files servers are controlled from a single entity.
- **Disadvantages for Centralized Storage**
  - A single storage solution has geographical barriers and this inflexibility requires all server storage to be located in one geographical location that now has to provide high-speed access to other sites
  - The initial costs for setting up a large centralized data store is actually significant, if it has to take care of an entire enterprise

### Logical Design Choice 2: Distributed Storage

The distributed storage model dictates that each hardware server and associated software service provides, configures and manages their own storage for the organization.

- **Advantages for Distributed Storage**
  - Distributed storage option provides the flexibility for each application to work in its own segmented environment and gives the application owner control of the data storage and design
  - Distributed storage also places minimum impact on other services by separating storage requirements from other application data concerns
- **Disadvantages for Distributed Storage**
  - There are many storage decisions to make, this means that for a project which requires its own storage solution, it will need to have expertise in storage configurations

  - This is a more expensive total cost of ownership (TCO), since the storage is distributed into different networks, and this leads to management overhead for monitoring and backup and recovery options

## Logical Design Choice 3: Hybrid Storage

The distributed storage model dictates that each hardware server and associated software service provides, configures and manages their own storage for the organization.

- **Advantages for Hybrid Storage**
  - The hybrid storage model provides the advantage that it provides flexibility based on creating a number of different storage solutions for applications and projects
  - Hybrid storage also allows the engagement of legacy storage solutions that has already been part of the enterprise
- **Disadvantages for Hybrid Storage**
  - A hybrid storage solution can be the most complex, and for a large enterprise this increases the management overhead along with total cost of ownership. A hybrid model also needs a firm set of requirements to be implemented from all of the business stakeholders.
  - Because of the added complexity, a hybrid solution may frequently be more expensive. This because it contains storage hardware and software from a variety of vendors.

Each of these logical storage design choices defines a storage pattern that can be used as a starting point. These design choices can provide guidance on how storage should be integrated within the organization. Storage architecture and design should also be communicated within the enterprise architecture for the organization. The optimal activity would be if business users, technology implementers, and facility managers can all communicate and agree on common storage solutions for the enterprise.

For your enterprise, if you follow a structured design process, you may want to conduct the following set of activities:

- Determine the storage system requirements
- Choose the appropriate storage technologies for the system
- Define the level of service and fault tolerance that the new storage system provides
- Define enterprise-level backup and recovery processes and procedures
- Assess how the new storage system affects your network and communications

# ENTERPRISE STORAGE MANAGEMENT ACTIVITIES

Here we want to discuss how to effectively provide storage management services for the enterprise. As we said earlier, managing distributed storage environments across the country is a crucial activity for large corporate organizations along with DoD and military organizations.

*Table 1. Network Dependencies for Storage Architecture*

| Dependency Name | Specific Requirements |
|---|---|
| Directory Service | A directory service is required to manage user authentication and authorization and manage other resource objects in the environment. |
| Backup and Recovery Services | Backup and recovery services are required for data availability and storage management. |
| Server Deployment | All SAN-connected servers must be installed with correct software driver and firmware |
| Network Devices | NAS/SAN client access occurs over an Ethernet network |
| Network Services | TCP/IP name resolution services are required for all hosts and storage management devices to communicate properly. Additional NAS/SAN management usually occurs through HTTP, TELNET, and/or Simple Network Management Protocol (SNMP) interfaces. |
| Infrastructure Management Services | Infrastructure management services are required to provide the resources for remote administration and support the storage systems in the enterprise environment. |

## Questions to Consider for the Enterprise

For a large enterprise, the following set of questions, in these particular topics, may help to establish the business need and determine how a new storage system can be managed.

- **Availability**
  - Which set of business applications are critical and which need to be safeguarded against unplanned outages?
  - The implementation would be different based on if it is an e-mail system, an enterprise software application, or an enterprise database.
- **Scalability**
  - What is the storage volume and capacity needed for the next several years?
  - The new storage solution should be able to grow not only in disk capacity but also for newer operating systems and new interfaces.
- **Budget**
  - What is the funding that has been allocated for storage purchases and the sustainment of the new storage solution?
  - This will determine the planning and the people resources necessary to accomplish the task.
- **Performance**
  - What are the peak performance values needed for key applications?
  - This will need to be obtained from business and application owners.
- **Data Backup**
  - What are the enterprise backup requirements for backing up all data types?
  - Is there a firm window for nightly backup and is there a service level agreement in place?
  - Should off-site backup facilities be necessary away from the LAN or WAN?
- **Server and Storage Utilization**
  - How much available but inaccessible storage could be used by data intensive applications for sharing throughout the organization?
  - Are there server and storage consolidation requirements within the data center for your enterprise?

## Net-Centric Challenges

Net-centric systems within the theater or battlefield pose a greater set of issues than the normal set of questions to ask when implementing a new storage solution. In line with military upcoming activities, a net-centric computing system should provide the following features:

- Prepare advanced computing capabilities while reducing space, weight and power requirements
- Allow integration with a wide variety of hardware platforms, including tracked, wheeled, and airborne vehicles
- Ensure that systems have a completely sealed, conduction cooled chassis providing protection against vehicle shock, vibration, dust, sand, salt, water, and fog
- Ensure that the system has an easy-to-remove, small form factor that is weather resistant and reliable under extreme conditions
- Prepare to build all hardware system interfaces in a rugged and environmentally-sound manner to ensure real-time operations at all times

## People and Process

As part of enterprise storage management activities, it is important to review the necessary technical personnel along with management processes to be followed. Two roles that have been defined by industry for the storage enterprise includes the Storage Manager, and the Storage Administrator. People who perform these roles are responsible for providing all of the data storage functions to the organization. It is also important to identify skilled specialists who can focus on the performance of the production systems and the tasks that need to be run on a daily basis. Backup and recovery activities may require separate personnel as dictated by organization's policies and procedures.

For a new enterprise storage solution, it is very important to put in place a codified set of information processes and policies to support and maintain the system on an ongoing basis. These processes fall into the following basic groups:

- **Data Protection**: These processes provide software routines that are used to proactively protect the data on the storage solutions.
- **Process Backout**: These processes ensure that all storage procedures have a process to return to good data in a known state.
- **Data Recovery**: These processes provide routines for recovering data in the event of a storage failure or data corruption.

## Dependency on Network and Communications

The primary dependency for the storage enterprise is the direct attachment to the network and communications infrastructure. Based on the different storage options, NAS and SAN both are totally dependent on the network in use within the organization, along with external and remote facility connections. The dependency also exists for DAS, since DAS invariably requires use of the network to provide backup solutions.

Storage management and storage services need to be coordinated with network personnel, network management capacity issues, and network management activities. A number of specific services that correlate with network and communications are required to implement a storage architecture for the enterprise. These services are listed in Table 1 along with specific requirements for the network.

The network architecture has to accommodate storage and archiving concerns along with data security concerns. In a distributed or a hybrid storage model, data is dependent on the network architecture, since data needs to be securely spread across many systems and locations. Storage services and hardware deployment need to ensure that the primary data file along with the metadata that describes the data is securely stored in network file servers. Network performance and quick storage and retrieval from computing devices are essential to providing an agile infrastructure for the future.

## REVIEW OF CHAPTER GOALS

The goals of this chapter were to:

1. Acknowledge that there is an ever greater need for digital storage based on today's technology:
   - Recognizer how storage size and volumes are rising exponentially in today's organizations.
2. How to engage in determining enterprise storage requirements:
   - Examine information technology infrastructure needs along with facility and personnel resource demands.
3. Understand the policy-based approach for storage life cycle management:
   - Address concepts such as ILM, HSM, and DLM.
4. Know the various information management stages for data storage:
   - Address data classification, data policy, data management, data classification, and tiered infrastructure.
5. Review storage architecture hardware designs and configurations:
   - Use of direct attached storage, network attached storage, and storage area networks.
   - Discuss benefits of centralized storage, distributed storage and hybrid storage.
6. Address enterprise storage management activities and challenges:
   - Address storage issues for large enterprises, net-centric challenges and dependency on network and communications.

## REFERENCES

Couture, A., (2005). *Storage Service Market Leaders Hold Their Ground.* Gartner Publication GG 06-28-05

DiCenzo, C., et al. (2005). *Hype Cycle for Storage Software Technologies.* Gartner Publication, GG 06-03-05

Ghosh, S. (2006, July). *Electronic Records Archive (ERA) Technology Evolution Plan.* National Archives and Records Administration, Lockheed Martin, 2nd ERA Technology Evolution Council Meeting.

Microsoft Corp. (2005, March). *Storage Enterprise Design*. Retrieved from www.microsoft.com

Office of the Assistant Secretary of Defense for Networks and Information Integration, Department of Defense Chief Information Officer (2007, April). *Directive 8320.02-P: Data Sharing in a Net-Centric Department of Defense.*

Rinnen, P., (2004). *Market Share: Network-Attached and Unified Storage, Worldwide.* Gartner Publication, GG-04-14-05

Thibodeau, K. (1997). Boundaries and Transformations: An Object-Oriented Strategy for Preservation of Electronic Records. European Commission. In *INSAR Supplement II: the Proceedings of the DLM-Forum on Electronic Records*. Luxembourg: Office for Official Publications of the European Communities.

# Section 4
# Assessing Net-Centricity in Organizations

*This final section discusses a number of aspects for assessing Net Centricity within organizations that also includes reviewing enterprise architecture, interoperability, and technology evolution. The last chapter provides an understanding of the upcoming target state of a service-oriented enterprise after the transformation. There are four chapters within this section.*

- ***Chapter 14:*** *Architecture Assessment of the Federal Enterprise Level*
- ***Chapter 15:*** *Net-Centric Assessment and Interoperability Testing*
- ***Chapter 16:*** *Technology Evolution Assessment for the Future*
- ***Chapter 17:*** *Achieving a Net-Centric Service-Oriented Enterprise*

# Chapter 14
# Architecture Assessment at the Federal Enterprise Level

## CHAPTER CONTENT

As you explore Chapter 14, it will cover the following topics:

- Overview of Federal Enterprise Architecture
- Federal Enterprise Architecture Reference Models
- Defining an Enterprise Architecture Transition Strategy
- Enterprise Architecture Assessment of Federal Agencies
- Identifying Maturity of an Agency Enterprise Architecture

## CHAPTER FOCUS

This chapter focuses on assessing the maturity of enterprise architecture within our federal government, which is peripherally tied to the net-centric readiness of military and commercial organizations. We provide an overview of federal enterprise architecture guidance and federal reference models to comply with

DOI: 10.4018/978-1-60566-854-3.ch014

OMB mandates. We define an enterprise architecture transition strategy that allows organizations to move from their current state to a target state. We then go ahead and assess federal agencies based on Clinger-Cohen Act and OMB mandates. We end with discussing enterprise architecture maturity and how to achieve it within a large organization.

The purpose of this final section is to focus on the assessment of technology architecture throughout the government and commercial enterprise. This chapter provides a description on the use of enterprise architecture to assess our federal government agencies. Based on federal mandates, enterprise architecture processes have proliferated within the government, and this allows the ability to adhere to new upcoming technologies such as the net-centric concepts proliferation by the defense department and military organizations.

## OVERVIEW OF FEDERAL ENTERPRISE ARCHITECTURE

Within the federal government, the President's Office of Management & Budget (OMB) ensures that technology advances are properly addressed within each of the federal agency budgets. As the overall budget stakeholder of the federal government, the OMB asks that each agency provide a technology roadmap ties that into the overall goals of the government to its public citizens.

A number of years ago, the OMB established the Federal Enterprise Architecture (FEA) program that enables government decision makers with a comprehensive business-driven blueprint of the entire Federal government. The FEA effort allows the creation of a common playbook that focuses on agency IT investments, and enhances collaboration among federal agencies. The higher-level goal is to drive toward a citizen-centered, results-oriented, and market-based organization as set forth within the President's Management Agenda (EOP FEA Consolidated Reference Model, 2007).

According to the OMB FEA guidelines, the core principles of this leading effort are:

- **Business-Driven:** The OMB recognizes that the FEA is most useful when it is closely aligned with government functional stakeholders – these include personnel who are crafting the agency strategic plans, mission statements, and yearly budgetary initiatives
- **Proactive and Collaborative across the Federal Government:** As agencies fully prepare their enterprise architecture, it is important to understand their commonality with other government agencies and that they proactively participate in information sharing. The FEA actively promotes the development, evolution and adoption of a common enterprise architecture framework for use by all government officials
- **Improve the Efficiency and Effectiveness of Government Resources:** The direction is to acknowledge that enterprise architecture development is an integral part of the capital investment process, and that no IT investment should be made without a business-approved architecture.

In implementing EA across the federal government, the FEA has developed a set of interrelated EA "Reference Models". These models have been promoted via the Federal CIO Council to each government agency along with federal contractors, vendors, and corporate organizations.

*Figure 1. Federal Enterprise Architecture Reference Models*

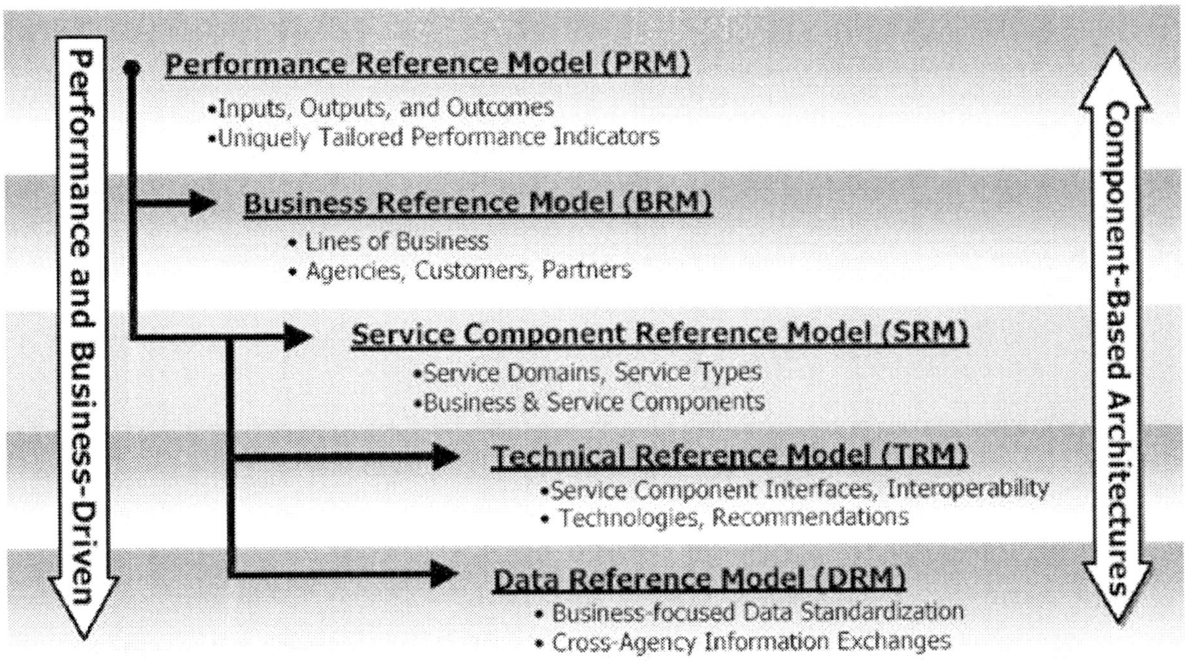


## FEDERAL ENTERPRISE ARCHITECTURE REFERENCE MODELS

The FEA Reference Models are designed to facilitate at the Congressional level, the analysis to identify duplicative investments, technology direction, information gaps, and opportunities for collaboration and sharing among federal agencies. The OMB has also offered an overall enterprise architecture framework called the Federal Enterprise Architecture Framework (FEAF) that works to enhance the maturity of enterprise development efforts within the government. The FEAF allows the use of a common vocabulary and the ability to gauge IT investments in a common and consistent manner. The OMB has provided for wide public use, the following five FEA Reference Models:

- **Performance Reference Model (PRM)**
- **Business Reference Model (BRM)**
- **Service Component Reference Model (SRM)**
- **Technical Reference Model (TRM)**
- **Data Reference Model (DRM)**

The following Figure 1 provides simple illustration of the purpose of these reference models and their inter-relationships.

Based on this illustration, we will be ensuring that the commonality in architecture is performance and business-driven, starting with the Performance Reference Model (PRM), then the

Business Reference Model (BRM), the Service component Reference Model (SRM), the Technical standards Reference Model (TRM), and the Data Reference Model (DRM).

## Performance Reference Model (PRM)

The PRM has been created as a framework that can be given to government agency personnel for performance measurement so that a common output can be achieved. This apples-to-apples comparison allows agencies to better manage the business of government at a strategic level. The PRM guidelines allow an agency's Enterprise Architecture to measure the success of their technology investments and the ensuing impact on strategic outcomes.

The PRM establishes common rules by which agency architecture is able to document its outputs and scores their program and business objectives. The PRM links the internal agency business processes to end-user customer-centric outputs. This allows proper resource allocation, and the evaluation of programs to figure out which ones are efficient and effective. The PRM focuses on the following objectives:

- Inputs, Outputs and Outcomes to improve the government's ability for strategic and daily decision making
- Set a clearly delineated "Line of Sight" measurement to better articulate the agency's goals and outcomes
- Address ongoing process and performance improvement so that government activities are clearly focused on their mission to provide better services for their citizens

## Business Reference Model (BRM)

The BRM focuses on the government functional stakeholders, and provides a framework that facilitates the ability to create a hierarchy of needs in alignment with the agency mission and goals. The FEA program has worked with the inter-governmental agencies to establish an overall scope and purpose of the federal government based on Lines of Business (LoB). These business lines address the overall government function to provide critical services to citizens and internal government operations.

The idea of the BRM is to describe the federal government around common business areas, and not along disparate agency-by-agency views. This then promotes inter-agency collaboration and goes toward the goals laid out by the Clinger-Cohen Act, along with the E-Gov and the Presidential Management Agenda.

The BRM allows all organizations to create their organizational hierarchy based on business and operational needs. The BRM utility serves as a model for an improved way of thinking as to how provide services to the end users. The BRM promotes the President's Management Agenda by ensuring that E-Government goals set forth are followed by each agency in a progressive manner. The agency CIOs can then align their business-focused enterprise architecture and their management processes with the budgetary guidelines of the OMB.

## Service Component Reference Model (SRM)

The SRM is a reference model that cascades down from the PRM and BRM, and focuses on the specific services that government provides on an ongoing basis. The SRM is a BRM-driven functional framework that classifies the overall Services into Service Components based on how they support business and performance objectives. The SRM identifies and classifies Service Components as Service Domains and Service Types that can be correlated with all of the information investments and assets supported by the Federal government.

The SRM model creates a directory listing of all the types of services provided by a government agency, and figures out common services that works throughout the federal space. This enables the reuse of business components and provides common service capabilities across the federal government. The SRM model also lends itself toward new IT architecture patterns such as Service-Oriented Architecture, where IT investments can be categorized as either service providers or service consumers.

The SRM directory of services is organized across horizontal service areas, and is independent of the Line of Business and business functions mentioned within the BRM. The Federal CIO Council has been working with government stakeholders and industry and corporate leadership to be able to leverage existing service initiatives, IT infrastructure activities, and government operations to provide the proper set of services for each government business instance.

## Technical Reference Model (TRM)

Since the PRM, BRM, and SRM concentrate primarily on the functional needs of the government, the TRM provides the technology standards to enable these other reference models. The TRM is component-driven, and provides a consistent framework for categorizing technology standards and upcoming information technologies. This then directly supports the individual Services, and enables the delivery of Service Components within the SRM.

Within the government arena, it is important that technology standards are common, and can be supported by all agencies. The TRM promotes E-Gov guidance and provides a foundation for advancing the reuse and standardization of technology. The Federal CIO Council and the OMB further believe that if each government agency aligns their capital investments with the TRM technical standards, then this allows agencies to leverage a common, standardized vocabulary with regard to technical interoperability. With regard to the DoD and Net-Centricity, the FEA TRM plays a central role as a common mandate to follow for OMB budgetary guidelines.

## Data Reference Model (DRM)

Among each of the reference models, the DRM is undoubtedly the most complex. This is because as more details of the agency enterprise architecture is discussed in public and information is shared, the underlying data is probably the most difficult to decipher. As a result, the Federal CIO Council decided to create a DRM that is a flexible and follows a standards-based framework to enable information sharing and reuse across the federal government. The DRM aspires to a distributed data framework that highlights common methods for description and discovery of common data and promotes a uniform level of data management practices.

*Figure 2. Baseline to Target Enterprise Architecture*

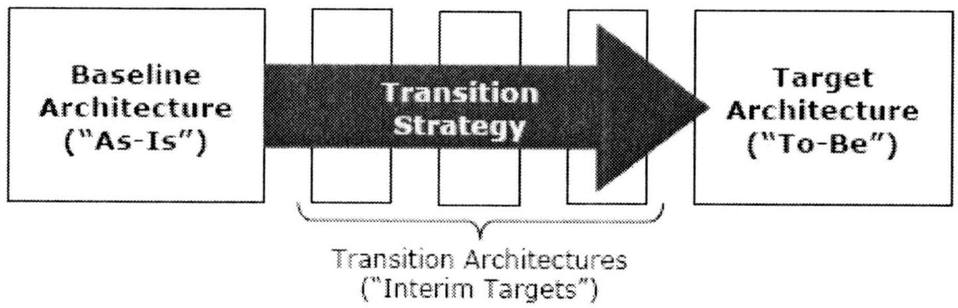


The DRM asks that government agency CIOs establish standard mechanism by which data is described, categorized, and shared.

According to the DRM guidelines, there are three standardization areas:

- **Data Description:** Describes data in a uniform manner so that there is agency level support for discovery and sharing.
- **Data Context:** Each set of functional data and their associated metadata is categorized based on taxonomies. The authoritative data source is identified along with proper usage and data security, and the support for transmitting this data to external stakeholders.
- **Data Sharing:** The ability to access and exchange of data based on Ad-hoc query requests and exchange of fixed transactions an ongoing basis between parties.

## DEFINING AN ENTERPRISE ARCHITECTURE TRANSITION STRATEGY

Now that we have described the Federal Enterprise Architecture Reference Models in detail, we would like to describe the critical component of EA, and that is an EA transition strategy. The creation of a transition strategy is recommended by the OMB for all federal agencies including the DoD. The transition strategy describes the overall plan for an organization to move from its current state to achieve its target state, or a "to-be" architecture within a specified timeframe. It tries to link the organization's technology investments to the target architecture. It also helps define the logical dependencies between transition activities such as programs and projects and the organization's investment priorities.

In the case of a net-centric transformation, an EA transition strategy is required that moves from the command and control hierarchy that exists within large organization to the agile, networked organization that is exemplified by a net-centric organization.

Figure 2 provides a diagram of an organization moving from the baseline "as-is" architecture to a target "to-be" architecture, based on a transition strategy that happens in three interim target states.

In order to prepare a transition strategy, both the baseline architecture and the target architecture needs to be documented. This would include a set of EA diagrams or roadmaps that is placed in a common repository accessible to required personnel within the organization. For a large organization, an EA transition strategy should address a multi-year timeframe, typically three to five years, for which the target architecture should be defined. The baseline architecture should have significant detail and

*Figure 3. Conceptual Enterprise Sequencing Plan*

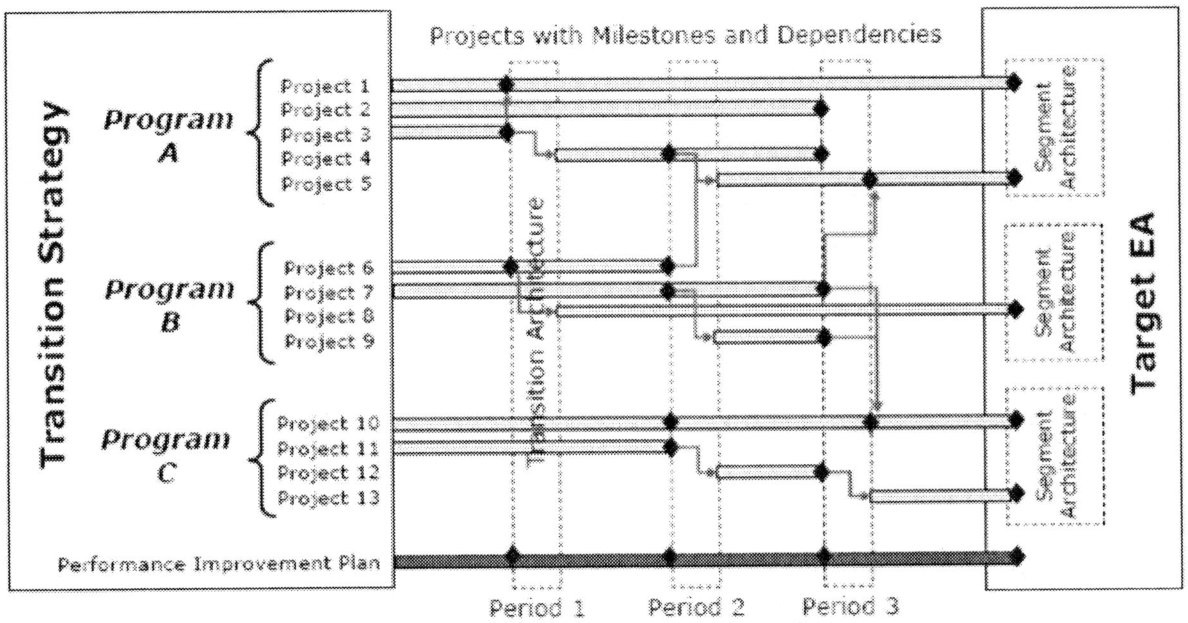


completeness so that it can be a starting point for the organization's transition strategy. The transition strategy should also be update as an organization's baseline and target Eas are updated on a periodic basis.

## Content for an EA Transition Strategy

The content for an EA transition strategy should contain the organization's defined programs and projects, and an evaluation of the gap analysis that is necessary moving forward.

- **Defined Programs and Projects** – The organization's projects and programs should be defined within the transition strategy and should be examined based on the context of the overall architecture. Projects and programs should feed directly into the investment management process, and should provide links as to if the project should receive its appropriated budget.
- **Redundancy and Gap Analysis** – The purpose of performing gap analysis and looking at redundancy and is to identify opportunities for consolidation or reuse in the baseline architecture and to identify "gaps" between the baseline and target architectures.

## Enterprise Sequencing Plan

As part of the transition strategy, another key item is to develop an enterprise sequencing plan as to how this transition will be conducted. The enterprise sequencing plan should provide an organization-wide view of the programs and projects across the board, and give leadership visibility so that the EA can be used for organization-wide planning.

Figure 3 provides an illustration of the sequencing plan that involves different programs within the organization on the left side and the sequential steps necessary to complete the development of these programs to achieve the target state. The enterprise sequencing plan can be referred to as a roadmap that is used by management to plan and track the progress of their program activities.

Here, the key elements of the sequencing plan are defined as follows;

- **Program** – A program is an activity or set of activities intended to help achieve a particular outcome for the end user. Within the Federal government, a program may need to be recognized by the Executive Branch and the Congress when making budgetary decisions.
- **Project** – This is a discrete, planned effort to achieve a specific goal or result within a brief timeframe. A program manager is accountable for each project as it moves through the investment process and implementation.
- **Project Interactions** – Interactions between projects should show dependencies between the programs and projects within the organization. An enterprise sequencing plan is not intended to replace ongoing project management or financial management activities. It is merely for illustrating on an overall basis, the external interactions.
- **Dependencies between Programs and Projects** – This illustrates visually the dependencies between programs and projects so the effects of budget decisions or slipping schedules can be quickly assessed by management. This would include impacts on performance milestones and plans to achieve the target state.
- **Target EA** – The target EA represents the future vision for the organization and serves as a blueprint for future conduct. The target EA should already be known before a transition strategy and should serve as an endpoint for the transition strategy.
- **Segment Architecture** – For federal agencies such as the DoD, a segment architecture is the technology architecture for an individual line of business (LoB) or a particular service component as described within the FEA Service Component Reference Model (SRM). A segment architecture adds more detail to the agency EA in a federated fashion.
- **Performance Improvement Plan** – Provides a summary of the performance goals and planned results from each project or program identified in the sequencing plan. A performance improvement plan also contains a consolidated view of the cost reduction and performance improvement goals, including how to get to interim milestones that is necessary to achieve the Target EA.

## ENTERPRISE ARCHITECTURE ASSESSMENT OF FEDERAL AGENCIES

This section provides a mechanism for assessing federal government agencies and subordinate large organizations. The assessment is based on the use of the Clinger-Cohen Act and the OMB EA Assessment Framework.

### Assessment Based on Clinger-Cohen Act Compliance

The Information Technology Management Reform Act of 1996, otherwise known as the Clinger-Cohen Act provides a way for federal agencies to assess their information technology investments. Instead of viewing information technology as a customized mix of technology specifications and network and communications infrastructure, the key element is to "improved operations" and "reduced costs."

Assessment questions include the following:

- How is IT helping to improve the Agency mission?
- What value am I getting for my IT investments?
- How can I reduce ineffective information systems that resulting in waste, fraud, and abuse?
- How can I pay better attention to the business processes necessary to conduct my critical operations?
- How can I reduce long standing, systematic problems within my IT infrastructure?

The following provides a full set of assessment parameters that allow current information organizations to transform to their future state. Adhering to these parameters can be regarded as best practices, for government agencies and agency contractors to consider:

- **Policy and Organizational**
    - Department/Agency missions, organization, functions, policies, procedures
    - Governing laws and regulations
    - Federal government decision-making, policy making process and budget formulation and execution process
    - Linkages and interrelationships among Agency Heads, COO, CIO, and CFO functions
    - Intergovernmental programs, policies, and processes
    - Privacy and security
    - Information management
- **Leadership/Managerial**
    - Defining roles, skill sets, and responsibilities of Senior Officials, CIO staff and stakeholders
    - Methods for building federal IT management and technical staff expertise
    - Competency testing – standards, certification, and performance assessment
    - Partnership/team-building techniques
    - Personnel performance management techniques
    - Principles and practices of knowledge management
    - Practices which attract and retain qualified IT personnel
- **Process/Change Management**
    - Techniques/models of organizational development and change
    - Techniques and models of process management and control
    - Modeling and simulation tools and methods
    - Quality improvement models and methods
    - Business process redesign/reengineering models and methods
- **Information Resources Strategy and Planning**
    - IT baseline assessment analysis
    - Interdepartmental, inter-agency IT functional analysis
    - IT planning methodologies

  - Contingency planning
  - Monitoring and evaluation methods and techniques
- **IT Performance Assessment: Models and Methods**
  - GPRA and IT: Measuring the business value of IT, and customer satisfaction
  - Monitoring and measuring new system development: When and how to "pull the plug" on systems
  - Measuring IT success: practical and impractical approaches
  - Processes and tools for creating, administering, and analyzing survey questionnaires
  - Techniques for defining and selecting effective performance measures
  - Examples of and criteria for performance evaluation
  - Managing IT reviews and oversight processes
- **Project/Program Management**
  - Project scope/requirements management
  - Project integration management
  - Project time/cost/performance management
  - Project quality management
  - Project risk management
  - Project procurement management
  - System life cycle management
  - Software development
- **Capital Planning and Investment Assessment**
  - Best practices
  - Cost benefit, economic, and risk analysis
  - Risk management- models and methods
  - Weighing benefits of alternative IT investments
  - Capital investment analysis- models and methods
  - Business case analysis
  - Integrating performance with mission and budget process
  - Investment review process
  - Intergovernmental, Federal, State, and Local Projects
- **Acquisition**
  - Alternative functional approaches (necessity, government, IT) analysis
  - Alternative acquisition models
  - Streamlined acquisition methodologies
  - Post-award IT contract management models and methods, including past performance evaluation
  - IT acquisition best practices
- **E-Government/Electronic Business/Electronic Commerce**
  - Strategic business issues & changes w/advent of E-Gov/EB/EC
  - Web development strategies
  - Industry standards and practices for communications
  - Channel issues (supply chains)
  - Dynamic pricing
  - Consumer/citizen information services
  - Social issues

*Figure 4. OMB EA Score Card*

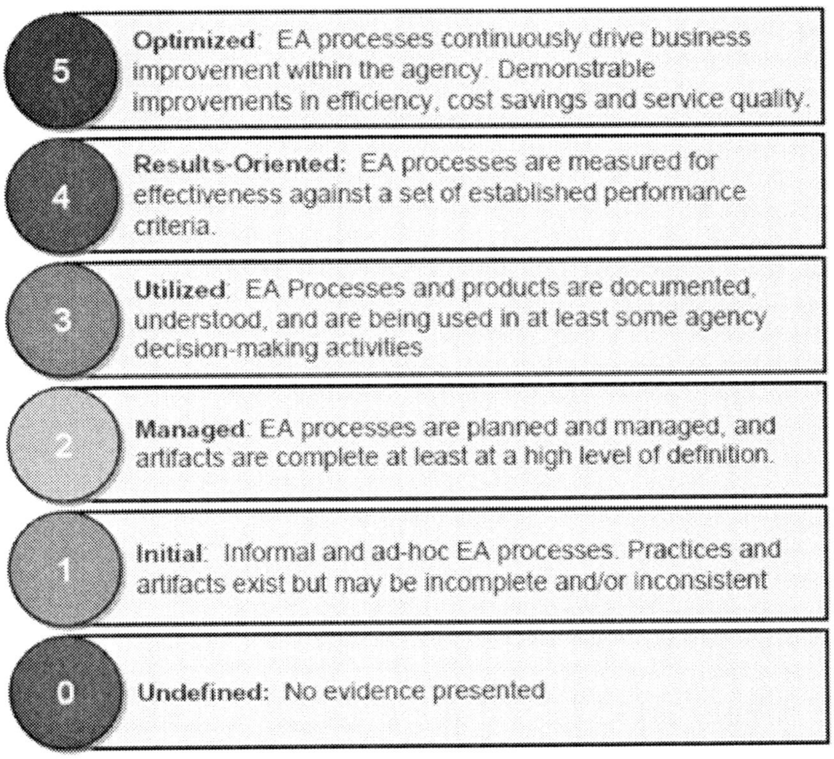


- **IT Security/Information Assurance**
  - Fundamental principles and best practices in IA
  - Threats and vulnerabilities to IT systems
  - Legal and policy issues for management and end users
  - Sources for IT security assistance
  - Standard operating procedures for reacting to intrusions/misuse of federal IT systems
- **Enterprise Architecture**
  - Enterprise architecture functions and governance
  - Key enterprise architecture concepts
  - Enterprise architecture development and maintenance
  - Use of enterprise architecture in IT investment decision making
  - Interpretation of enterprise architecture models and artifacts
  - Data management
  - Performance measurement for enterprise architecture
- **Technical Activities**
  - Emerging/developing technologies
  - Information delivery technology (Internet, Intranet, kiosks, etc.)
  - Desk Top Technology Tools

## Assessment Based on OMB EA Framework

The OMB uses an EA Assessment Framework (OMB Framework, 2008) to assess each federal agency and subordinate organization. These assessment activities also have credence in the commercial marketplace for large corporate firms who are also organizing their business areas. Here are a set of assessment indicators for senior leadership within the organization, and for reporting purposes to external entities.

- **Target Enterprise Architecture and Enterprise Transition Plan**
  - Measures how effectively and efficiently the Target EA is addressing gaps, redundancies, and costs in the IT portfolio and environment.
  - Government agencies must submit a comprehensive Target EA.
  - Addressing how underperforming programs would be changed based upon performance assessments.
  - The enterprise transition plan needs to depict how the organization would head towards the target state.
- **Architectural Prioritization**
  - Measures how the development of agency segment architectures are aligned to high priority needs
  - Agencies should have a structured process to link agency strategic planning and performance management processes
  - Determine high priority organization activities and demonstrate how each activity fits the purpose of the agency services
- **Scope of Completion**
  - Calculates the percentage of the organization's enterprise IT portfolio funding that is provide by completed segment architectures
- **Internet Protocol Version 6 (IPv6)**
  - Ensure that the organization's EA incorporates Internet protocol version 6 (IPv6) network transformations into the IT infrastructure segment architecture and IT investment portfolio.

These assessment indicators are then used to prepare performance improvement plans, enterprise transition plans, and processes and outcomes. When organizations prepare their portfolio of business and information capabilities, new investments need to be tracked based on the architecture assessment parameters. Collaboration and reuse of information technology and infrastructure is encouraged so that it increases flexibility in enhanced capabilities. A governance structure has to be in place to ensure that assessment tasks are carried out properly and with the requisite personnel oversight. It is also important to receive executive-level sponsorship on an ongoing basis to effectively carry out change management and deployment processes.

## IDENTIFYING MATURITY OF AN AGENCY ENTERPRISE ARCHITECTURE

For federal government agencies, the OMB conducts an annual, comprehensive assessment of the agency's enterprise architecture. The OMB expects agencies such as the DoD to submit quarterly progress reports on their success in achieving the milestones set forth in their transition strategy and performance improvements they have realized from setting up an agency enterprise architecture.

The annual assessment process is intended to provide a comprehensive review of the agency's enterprise architecture program and management controls. The organizational review focuses on the following elements:

- Completion of an enterprise architecture along with the related artifacts
- Use of enterprise architecture to drive improved decision-making
- Results achieved to improve the agency's effectiveness

Figure 4 provides a scale or score card for judging the maturity status of the agency's enterprise architecture.

This is a vertical scale that rates a 0, 1, 2, 3, 4, or 5 based on how well enterprise architecture is institutionalized within the agency. Here is a categorization of the levels:

- The first or Level 0 says that there is no evidence of enterprise architecture present.
- Level 1 rates the agency at the initial level, where efforts have just started to progress.
- Level 2 states that the agency has now achieved a managed level that allows personnel to manage their baseline and target state.
- Level 3 now says that the agency is starting to utilize the architecture efforts
- Level 4 and Level 5 focus on making the enterprise architecture results-oriented and optimized. The idea is to be better able to navigate toward a defined target state.

## REVIEW OF CHAPTER GOALS

The goals of this chapter were to address:

1. What are the dramatic changes brought out by the Clinger-Cohen Act with regard to information technology?
    - Describe how it helped federal agencies in getting better value for their IT investments
2. What are the set of Clinger-Cohen Competencies?
    - Address how they provide a set of organizational compliance activities that agency considers on an annual basis as part of the budgetary process
3. How does the Federal Enterprise Architecture help in facilitating information technology investment decisions?
    - Describe the FEA Reference Models which means an understanding of:
        i. Performance Reference Model
        ii. Business Reference Model
        iii. Service Reference Model

    iv. Technical Reference Model
    v. Data Reference Model

4. How can we define an enterprise architecture strategy that allows organizations to move toward a target state?
   - Describe the process of producing a baseline architecture and target architecture
5. Develop an enterprise sequencing plan that addresses parallel program efforts
   - Put together a large organization level program chart that tracks efforts and evaluates program dependencies
6. Realize that the maturity of the enterprise architecture for a large organization requires a dedicated set of steps
   - Identify the 0 to 5 stepwise process to rate organizations in their enterprise architecture maturity

## REFERENCES

Commonwealth of Massachusetts Enterprise Information Technology Architecture, Enterprise Technical Reference Model – Version 3.5, (2005, September 21).

Department of Defense Business Transformation Agency. (2009). BEA Architecture Product Guide Version 6.0. Retrieved from http://www.bta.mil/products/bea.html

Executive Office of the President, (2005, December). *Federal Enterprise Architecture Program EA Assessment Framework 2.0.*

Executive Office of the President (2007, October). *FEA Consolidated Reference Model Document,* (Version 2.3). Washington, DC: Author.

Ghosh, S., Miller, C., Przysucha, J., & Tiemann, M. (2005, January). *Advancing Enterprise Architecture Maturity, Version 2.0.* Industry Advisory Council White Paper. Department of Defense (2007). *DoD Architecture Framework Version 1.5 - Volume I: Definitions and Guidelines.*

Office of Management and Budget (2008, December). *OMB Enterprise Architecture Assessment Framework, Version 3.0.*

Office of the Assistant Secretary of Defense for Networks and Information Integration, Department of Defense Chief Information Officer (2006, April). *Clinger-Cohen Act Compliance Certification of Major Automated Information System for Fiscal Year 2006.*

Spewak, S. H. (2002). *Enterprise Architecture Planning: Developing a Blueprint for Data, Applications, and Technology,* (2nd ed.). Chichester, UK: John Wiley & Sons.

# Chapter 15
# Net-Centric Assessment and Interoperability Testing

## CHAPTER CONTENT

As you explore Chapter 15, it will cover the following topics:

- Assessing Net-Centric Transition
- Net-Centric Data Assessment
- Net-Centric Services Assessment
- Net-Centric Information Assurance Assessment
- Communications and Transport Assessment
- Interoperability Testing for Net-Centric Development
- CASE STUDY: University of New Hampshire Inter Operability Laboratory (IOL)

DOI: 10.4018/978-1-60566-854-3.ch015

## CHAPTER FOCUS

This section provides you with the ability to assess net-centric and interoperability principles within organizations. This chapter provides a set of checklists that have been placed in tables for organizations to figure out how to evaluate net-centric data assessment, net-centric services assessment, information assurance assessment, and communications and transport assessment. It then provides an understanding of interoperability testing to promote net-centric development, which involves stages of testing and the use of SOA-based services. We end with a case study of the University of New Hampshire Interoperability Laboratory that has been a fixture in providing testing support for IPv6 and network testing.

## ASSESSING NET-CENTRIC TRANSITION

In this chapter, we now discuss the ability to assess, test, and evaluate a net-centric transition process for systems and organizations. These organizations may grow to include any type of entity, which includes federal government organizations, state and local government, large defense-related corporations, small business organizations, and universities and research consortia.

The four areas of assessing net-centric alignment include the following:

- Data Assessment
- Services Assessment
- Information Assurance Assessment
- Communications and Transport Assessment

We provide a set of questions that need to be answered by organization personnel as they decide how to proceed in transitioning their organizations to adopt net-centric principles. The questions have been from the net-centric checklist prepared by the DoD CIO organization (Net-Centric Checklist Version 2.1.4, 2007) to be answered by DoD Component organizations, Armed Force Services and other DoD organizations.

However, these questions are universal in their appeal and have been modified to address any organization that is currently trying to adopt a SOA-based architecture, a consumer-subscriber model, and a distributed information environment. This addresses any organization that includes large corporations, governmental organizations and numerous business entities.

It makes sense that all of the questions are not applicable to all organizations or all environments. In answering these questions, any person would realize this very quickly. However, the reason to review and capture these questions in totality is that it allows the entire set of information to make sense and work together. Since military organizations are larger and more complex that most commercial entities, the additional degree of rigor necessary to answer all of the questions creates an organization that has a greater degree of interoperability and reliability.

## NET-CENTRIC DATA ASSESSMENT

These set of questions is in line with the overall Net-Centric Data Strategy (DoD CIO NCDS, 2003) and provides a detailed set of questions to be answered when designing and developing a system.

*Table 1. Assessment to make data visible*

| | |
|---|---|
| **Make Data Visible** | 1. Describe how the system is aligned with the DoD Net-Centric Data Strategy. |
| | 2. Does the system provide discovery metadata, in accordance with the DoD Discovery Metadata Standard (DDMS), for all data posted to shared spaces? |
| | 3. Is all of the data that can and should be shared externally beyond the programmatic bounds of your system visible (i.e., advertised) to all potential consumers of the data? |
| | 4. Describe how the system is making its data assets visible to consumers. |
| | 5. Describe how consumers are able to locate the data assets available from your system. |
| | 6. Describe how the system is making use of Web service standards (e.g., SOAP [Simple Object Access Protocol], WSDL [Web Services Description Language], UDDI [Universal Description, Discovery and Integration]) to make its data assets visible. |
| | 7. Describe any subscribe/notify mechanisms for the visible data assets available within the program that alert users and other applications when data has been created or updated. |
| | 8. Describe where potential consumers can go to become aware of the data assets being made visible by your program. |
| | 9. Describe how the program provides dynamic, flexible, and threat-tailorable solutions for exchanging data assets between different security domains (i.e., cross-domain) with flexibility to accommodate new operational needs with minimal impact on system and mission performance. |
| | 10. Describe how data posted to shared spaces is controlled and managed by the applicable security policies or regulations and how these IA controls are enforced. |

- Table 1 provides a set of questions to make data visible.
- Table 2 indicates how to make data accessible.
- Table 3 provides questions as to how to make data understandable.
- Table 4 addresses how to make data trustable.
- Table 5 questions how to make data interoperable.
- Table 6 assesses how to provide data management.
- Table 7 assesses how to be responsive to user needs.
- Table 8 assesses how to obtain reliable and secure information.

## NET-CENTRIC SERVICES ASSESSMENT

Software applications in the form of Web Services provide an environment that allows organizations to perform rapid software development and deployment. These automated services codify a well-defined set of realizable capabilities that can be used with external organization's services that address business user needs.

- Table 9 asks a set of questions with regard to Service-Oriented Architecture.
- Table 10 addresses the need for open architecture.
- Table 11 provides questions on scalability.
- Table 12 provides questions on availability.
- Table 13 assesses how to accommodate heterogeneity.
- Table 14 provides questions on decentralized operations and management
- Table 15 assesses how to provide enterprise service management

*Table 2. Assessment to make data accessible*

| | |
|---|---|
| **Make Data Accessible** | 1. Is all of the data that can and should be shared externally beyond the programmatic bounds of your program accessible to all potential consumers of the data with sufficient access permissions and without any additional programming effort? |
| | 2. Describe for each visible data asset what the data consumer needs to access the data (e.g., an application client, a Web portal, access to a Web service, access to a shared data storage area, an XML (eXtensible Markup Language) schema/parser, etc.) |
| | 3. Are there any limitations for the client appliance (e.g., workstation, desktop, laptop, personal mobile devices) to access your data assets? |
| | 4. Has the program explicitly identified the potential universe of consumers of that data? (for example, local users, COI users, enterprise users) |
| | 5. Describe the program's architecture and the data separation from the presentation and business logic. |
| | 6. Describe the security mechanisms used to restrict access to specific, visible data assets. How will the associated metadata labels be used to support these security mechanisms? |
| | 7. What mechanisms are planned or implemented to protect the data in transit to the consumer? This would include protection from modification of the data, protection from unauthorized eavesdropping, or protection from data becoming "lost" in transit. |
| | 8. What mechanisms are planned or implemented to protect the data at rest within a consumer client? This would include protection from modification of the data, protection from unauthorized disclosure, or protection from data becoming "corrupted" or otherwise unavailable for mission use. |
| | 9. What mechanisms are planned or implemented to protect the data at rest within the service providers systems? This would include protection from modification of the data, protection from unauthorized eavesdropping, or protection from data becoming "corrupted" or otherwise unavailable for mission use. |
| | 10. Describe how the visible data assets are made available to other users outside the Community of Interest with a need for the data. |
| | 11. Describe the common design patterns employed in the program that aid in the accessibility of data assets. |
| | 12. Describe how the program provides assurance that there is timely and reliable access to data assets anytime, anywhere for authorized users/entities. Availability is a core IA function that is critical to ensuring successful mission execution. |
| | 13. Describe how access control and IA policy enforcement will be used to ensure that only authorized users/entities can access restricted data. |

*Table 3. Assessment to make data understandable*

| | |
|---|---|
| **Make Data Understandable** | 1. Is all of the data that can and should be shared externally beyond the programmatic bounds of your program sufficiently documented and understandable that any potential consumer can comprehend the structural and semantic meaning to determine how they may use it appropriately? |
| | 2. Is all of the data that can and should be shared externally beyond the programmatic bounds of your program sufficiently documented and understandable that any potential consumer can comprehend the structural and semantic meaning to determine if they can reliably use the metadata to make access control decisions on sensitive data? |
| | 3. Explain how the program is making use of the DoD Metadata Registry and Clearinghouse. |
| | 4. Describe the source of all XML elements. Has the DoD Metadata Registry been used whenever possible? Have newly defined XML elements been registered with the Registry? |
| | 5. Describe any data schemas or standards being applied in the program. |
| | 6. Describe any automated mechanisms that are available for data mediation/translation (e.g., XSL [eXtensible Stylesheet Language], XSD [XML Schema Definition]) |
| | 7. Describe any automated mechanism that enforces translation of security markings from one policy domain to another. |

*Table 4. Assessment to make data trustable*

| | |
|---|---|
| **Make Data Trustable** | 1. Can all potential consumers of all of the data available from your program determine the data pedigree (i.e., derivation and quality), security level, and access control level of your data? |
| | 2. Describe for each visible data asset in the program whether the program is the authoritative data source. |
| | 3. Describe what measures the program takes to ensure the integrity of the data (for internally used data, externally used data, and data that simply transits the program). |
| | 4. Describe what measures the program takes to ensure that the program data is only provided to consumers via authorized sources. |

*Table 5. Assessment to make data interoperable*

| | |
|---|---|
| **Make Data Interoperable** | 1. Does all of the data that can and should be shared externally beyond the programmatic bounds of your program have sufficient metadata descriptions and automated support to enable for mediation and translation of the data between interfaces? |
| | 2. Describe any programming changes that would need to be made to the program if a new consumer of a visible data asset were identified. |
| | 3. Identify the published net-centric interoperability standards (e.g., DDMS) to which the program adheres. |
| | 4. Describe the process a consumer would follow to a) request changes in the format (syntax or semantic) of the visible data asset; b) report a problem with a data asset; or c) request additional data from the data provider. |

*Table 6. Assessment to provide data management*

| | |
|---|---|
| **Provide Data Management** | 1. Is there sufficient management of all of the data available through your program to adequately maintain and improve your data assets within a changing environment? |
| | 2. Describe the effort associated with the program to define, develop, and maintain an ontology (i.e., schemas, thesauruses, vocabularies, key word lists, and taxonomies) that best reflects the community understanding of the "visible data assets." |
| | 3. Describe your processes for ensuring the usefulness and timely availability of all data assets associated with your program. |
| | 4. Describe the various data survivability scenarios considered in your program. |

*Table 7. Assessment to be responsive to user needs*

| | |
|---|---|
| **Be Responsive to User Needs** | 1. Are perspectives of users, whether data consumers or data producers, incorporated into data approaches via continual feedback to ensure satisfaction? |
| | 2. What tools, services, processes, and resources is the program providing to facilitate user feedback and program responsiveness with respect to data needs? |
| | 3. What metrics are being used to determine responsiveness to user data needs? |
| | 4. What is the degree of collaboration with respect to data that is enabled and is occurring among the user community (ies) and the program developers? |
| | 5. What are measured/assessed trends over time with respect to the program's responsiveness to user data needs and degree of satisfaction towards meeting those needs? |
| | 6. What are the programs plans to enhance responsiveness to user data needs? |

*Table 8. Assessment to obtain reliable secure information*

| | |
|---|---|
| **Obtain Reliable and Secure Information** | 1. Describe the protection mechanisms for program data to ensure that undetected compromises are contained and do not allow an adversary to access restricted or sensitive program data while still maintaining visibility to authorized users? |
| | 2. Describe the techniques that inhibit an adversary who has compromised a client or server from accessing all sensitive program data and services within the enterprise. |

*Table 9. Service Oriented Architecture assessment*

| | |
|---|---|
| **Service-Oriented Architecture** | 1. Describe how the program will make its unique services/applications available to the GIG community. |
| | 2. Describe who (e.g. which individual-types, COIs, or Roles within COIs) are expected to consume the services which your program provides? |
| | 3. Describe which services your program expects to consume (or utilize) as input to your service(s). Include who is expected to provide these services, and whether they are enterprise services, core services, COI-Specific services, or other type of service provider. |
| | 4. Does your program have an Operational-Connectivity Diagram (OV-2) that models how your services are provided and/or consumed by the other GIG Service providers/consumers? |
| | 5. Has you program identified the "operational threats" envisioned against your service-based operational nodes (OV-02) and/or need-lines between operational nodes? |
| | 6. Has your program identified specific "shared spaces" that will be used to facilitate the assured sharing of information as part of your service provision/consumption? |

*Table 10. Need for open architecture*

| | |
|---|---|
| **Need for Open Architecture** | 1. Is the system architecture based on loosely coupled interactions, enabling the internal components to map to well-defined external interfaces? Describe. |
| | 2. Are Web services implemented by the program built using the following core standards? |
| | 3. Does your organization abide by Internet or Web foundational standards as has been promoted by the IETF and W3C organizations? |
| | 4. Does your organization able to adopt emerging Web standards and best practices with regard to Web Services? |
| | 5. Does your organization accept and promote XML as a data messaging interface and all of the associated standards that are related with XML usage? |
| | 6. Does your organization accepts and promotes specific Web Services standards and is using the most trusted set of industry emerging standards? |

*Table 11. Ability to provide scalability*

| | |
|---|---|
| **Scalability** | 1. What estimates of service usage have been developed (How many expected consumers? How many service invocations per hour (or per some appropriate unit of time)? On which assumptions or empirical tests are these estimates based? |
| | 2. What performance analysis has been done to understand or predict the ability of the service to handle the expected load (e.g., number of end-user consumers or number of calling services)? |

*Table 12. Ability to provide availability*

| | |
|---|---|
| **Availability** | 1. Does the program have a Continuity Of Operations Plan (COOP)? |
| | 2. What is the plan for providing service during routine maintenance (hardware and software)? |
| | 3. What is the plan for providing service during catastrophic failures (e.g., massive outages of the power grid, physical destruction of the hosting facility)? |
| | 4. Does the continuity of operations plan include procedures for finding, housing, and supporting the technical support staff during an emergency action? |
| | 5. What threat scenarios have been considered in the planning? |
| | 6. What mechanisms are in place to detect denial or degradation of service attacks against GIG services and support the dynamic capability to contain, recover, restore, and reconstitute service? |
| | 7. What program features support delivery of program service while overcoming adversary attempts to deny service. |

*Table 13. Ability to accommodate heterogeneity*

| | |
|---|---|
| **Accommodate Heterogeneity** | 1. How will the service adapt to delivering end-to-end capabilities over a range of bandwidths (from low bit-rate tactical communications to multi-gigabit backbone service), environments (fixed and wireless), and end-user devices (e.g., PDAs, laptops, workstations, mainframes)? |
| | 2. Describe how the program will support delivery of capabilities to thin or browser-based clients, especially those that provide adaptability to a variety of disadvantaged 'edge' environments for end-users. |

*Table 14. Provide decentralized operations and management*

| | |
|---|---|
| **Decentralized Operations and Management** | 1. How will the program incorporate a single sign-on service to allow users anywhere on the Global Information Grid to obtain program services and data with a globally recognized set of identity/authorization credentials? What protection is there from compromised GIG Identities and the threat they represent within a single sign-on environment? |
| | 2. How will the program incorporate a privilege management system that allows assignment of privileges to access Global Information Grid services and allows acceptance of privileges granted from other Global Information Grid entities? |
| | 3. How will the program report enterprise management data to the Global Information Grid? How will the program interface and allow control of program service offerings and infrastructure by the Global Information Grid? |
| | 4. How will the program interface with the Global Information Grid to provide an enterprise wide Computer Network Defense (CND) while also providing CND services within the program? |

*Table 15. Provide enterprise service management*

| | |
|---|---|
| **Enterprise Service Management** | 1. Does the service provide instrumentation to enable the service provider to determine the current operational state and performance level of the service? |
| | 2. What protocols and standards are used to collect and disseminate service management information, and in what format will it be made available? |
| | 3. List and provide completed examples, including net-centric and NetOps elements/attributes, of all Service Level Agreements (SLA) negotiated by this service provider with others (e.g., transport layer programs, other service providers, consumers). |
| | 4. How will the program interface with the Global Information Grid to provide an enterprise wide Computer Network Defense (CND) while also providing CND services within the program? |

## NET-CENTRIC INFORMATION ASSURANCE ASSESSMENT

Since, DoD and military organizations have to abide by unique information assurance and security postures, the following set of questions are clearly specific to DoD organizations. However, based on reviewing applicable security regulations and directives, other organizations can modify these questions to address their own local needs.

- Table 16 – addresses net-centric information assurance posture for continuity of operations.
- Table 17 – addresses identity management, authentication and privileges.
- Table 18 – addresses privilege management.
- Table 19 – addresses how to mediate security assertions
- Table 20 – provides questions on cross security domains exchange

## COMMUNICATIONS AND TRANSPORT ASSESSMENT

As we have been saying all along – to provide a transformation as large in scope as the military, a very concerted effort needs to take place for a communications and transport infrastructure. For the defense sector, this is the foundational infrastructure is a foundation for the DoD, military and intelligence community. To realize the overall vision it is best to construct a communications transport infrastructure that:

- Follows the Internet Model
- Creates a comprehensive information grid from smaller component building blocks
- Designs component systems with interoperability, evolvability, and simplicity in mind

*Table 16. Net centric IA posture and continuity of operations*

| | |
|---|---|
| **Net Centric IA Posture and Continuity of Operations** | 1. Has a Mission Assurance Category and Confidentiality Level been identified for this program and is it documented in the capabilities document? |
| | 2. Are the assigned Mission Assurance Category and Confidentiality Level appropriate for this system in context with the Net Centric operational, systems architectures and technical standards within which it will function? |

*Table 17. Identity management, authentication and privileges*

| | |
|---|---|
| **Identity Management, Authentication and Privileges** | 1. How does the system use identity management for authentication (e.g., biometrics, Common Access Card, or passwords)? How does the program manage identity and privileges? |
| | 2. How are consumers assured that they are indeed receiving services of your program services from you (as an authorized source of these services) rather than from some other unauthorized service provider masquerading their identity? |
| | 3. What assurance does your program have that the services they consume within the GIG environment are from authorized sources and not from hostile service providers masquerading as authorized service providers? |
| | 4. How does the program provide authentication of its users in a fashion that will be accepted across the GIG? How will the program accept authentication credentials from users (including coalition partners) from anywhere across the GIG. |

*Table 18. Provide privilege management*

| | |
|---|---|
| **Privilege Management** | 1. Does the system provide different services (or different responses) based on the identity of the service requestor? |
| | 2. Does the system require that any specific criteria be satisfied before services or portions of services are delivered? |
| | 3. How does the program manage user privileges in a fashion that will be accepted across the GIG? How will the program accept privilege assertions from users (including coalition partners) from anywhere across the GIG. |

*Table 19. Mediate security assertions*

| | |
|---|---|
| **Mediate Security Assertions** | 1. Does the program mediate security assertions (to pass security related information between systems, processes, and domains)? If yes, how does the program mediate security assertions? |
| | 2. Does the program allow mediation of security assertions at policy enforcement points where the policy enforced at the policy enforcement point is managed and changeable by authorized users throughout the Global Information Grid? |

*Table 20. Cross security domains exchange*

| | |
|---|---|
| **Cross Security Domains Exchange** | 1) Has your program defined the Information Domains (or types of information domains) that establish the context for which your programs will be providing/consuming their associated services? |
| | 2) Has COI been defined to facilitate the appropriate information transactions within or between information domains? (e.g. "Are COIs explicitly defined to help manage & control information domain content?) |
| | 3) Has your program described (or modeled – e.g., OV-02 diagrams) the types of information transactions that are expected to occur within your respective information domains and between information domains? |
| | 4) Does the program need to exchange information or data elements across security domains (e.g., email, structured data sets, unstructured documents, imagery, etc.)? Include in this discussion connections to any existing/external systems with differing security policies. |
| | 5) How does the program accomplish the information or data exchange? Is this mechanism/capability inherent in the program or dependent upon some other program for this capability and if known, which program? |
| | 6) Does the system provide dynamic, flexible, and threat-tailorable solutions for exchanging information and services between different security domains (i.e., cross-domain) with flexibility to accommodate new operational needs with minimal impact on system and mission performance? |
| | 7) How does the system provide:<br>a. Protection against modification or destruction of information, services, or resources?<br>b. Assurance that information is not disclosed to unauthorized entities, services, or processes?<br>c. Assurance that the originator of and action (e.g., information process) and the recipient of a reaction are known (so that neither can deny having processed that action)?<br>d. Assurance that there is timely and reliable access to information, services, and resources anytime, anywhere for authorized users/entities? |
| | 8) Does the system have digital IA/security policy management capabilities that define, enforce, and manage the digital security policy rules, guidelines and standards for securely exchanging data and services across security domains? |
| | 9) Does the system provide a capability that enables security domains with different policy constructs to exchange information and be interoperable? |

The following questions are based on assessing the above general principles. The goal is wherever possible for systems to follow an Internet model" that allows "plug and play" characteristics to the overall network grid.

- Table 21 – assesses the organization's IPv6 compliance
- Table 22 – asks how to provide a packet switched infrastructure
- Table 23 – addresses layering and modularity
- Table 24 – addresses information transport goals
- Table 25 – assesses the organization's network connectivity
- Table 26 – ensures a concurrent transport of information flows
- Table 27 – assesses the different levels of quality of service
- Table 28 – addresses inter-network connectivity between organizations

## INTEROPERABILITY TESTING FOR NET-CENTRIC DEVELOPMENT

The implementation of web services allows the software industry to shift from a client-server model to a model where web based components are combined to build distributed applications.

In this net-centric assessment and testing, a web service is defined as any service that is accessible through the use of standard web protocols like Extensible Markup Language (XML) and Simple Object Access Protocol (SOAP). This also implies the use of facilitating specifications like Web Services Descriptive Language (WSDL) and Universal Description, Discovery, and Integration (UDDI) in order to specify the interface to the service.

The net-centric development environment quantifies the boundaries for the performance of web services. With regard to the testing process for software applications, performance testing has often focused on the overall application user interface. As the environment shifts toward testing web services, this standard for performance testing will no longer hold.

Although performance tools exist for directly testing the functions of web services, there are other considerations that must be considered.

- Ensure to test and evaluate all the other services and application servers that a web service calls in order to fulfill its function.
- These services and application servers affect the overall performance of any web service that calls them.
- Downtime from a piece of code that is part of a service's dependency will also cause downtime in that web service.
- Test and evaluate all external specifications for a service, which are essentially business processes that define the use of a service within an application environment.

*Table 21. Assess IPv6 compliance*

| | |
|---|---|
| **IPv6 Compliance** | 1. Describe the program's migration plan to an IPv6 environment. If not, when is it programmed to migrate? |
| | 2. What transition technologies such as tunneling, dual stack, etc., are being adopted? |
| | 3. Describe the transition approach in terms of network topology (e.g., what regions of the networks are IPv4 and what regions are IPv6). |

*Table 22. Provide a packet switched infrastructure*

| | |
|---|---|
| **Packet Switched Infrastructure** | 1. Describe any information flow between networks within your system or between your system and networks external to your system, which is not in the form of IP packets/datagrams. |
| | 2. Describe any information flow between networks within your system or to/from networks external to your system that does not pass through an IP router or layer-3 (IP) packet switch. |
| | 3. Describe the method(s) by which your system can accept IP datagrams from external networks that are destined for hosts within your system, and the ability of your system to act as a transit network for IP datagrams with an origin and destination that are external to your system. |

*Table 23. Layering and modularity*

| | |
|---|---|
| **Layering and Modularity** | 1. Describe all instances in your communications infrastructure where a logical or physical coupling or dependency exists between different layers of the protocol stack (e.g., if your system were to replace Ethernet with Token-Ring at layer 2, would the routing protocol you currently use at layer 3 fail or degrade?). |
| | 2. If you replace a current signal in space with a new physical layer signal, does your current layer 2 or layer 3 fail or degrade? |

*Table 24. Assessing transport goals*

| | |
|---|---|
| **Transport Goals** | 1. Is your organization headed toward a full convergence of traffic (voice, video, and data) on a single IP internet? Describe the process, protocols, and equipment used to provide full convergence of traffic (voice, video, and data) on a single IP inter network. |
| | 2. Describe any information flow between networks within your system or to/from networks external to your system that does not pass through an IP router or layer-3 (IP) packet switch. |
| | 3. What provisions have been made for increased redundancy using resource/path diversities since single link or node failure will take down three (voice, video and data) or more services at the same time? |
| | 4. If secure voice is part of the system, is it interoperable with other strategic and tactical secure voice systems with a 99% Threshold Key Performance Parameter? Is the secure voice system interoperable with coalition forces? |

*Table 25. Assessing network connectivity*

| | |
|---|---|
| **Network Connectivity** | 1. What link (layer-2) protocols will be used for transport of IP traffic? What standards are being used for these protocols? |
| | 2. What terminal or radio-to-network interfaces will be used for transport of IP traffic? What standards are being used for these interfaces? |
| | 3. What network-to-end user host interfaces will be used for transport of IP traffic? What standards are being used for these interfaces? |
| | 4. Does the system avoid any single point of failure by using multiple connectivity paths (not susceptible to the same threat) and media? |

*Table 26. Concurrent transport of information flows*

| | |
|---|---|
| **Concurrent Transport of information Flows** | 1) Describe the process (and protocols) that are used to provide convergence of traffic from multiple security domains on a single IP inter-network. |
| | 2) Describe the approach for providing an information infrastructure with a colorless or black core. |

*Table 27. Differentiated management of quality of service*

| | |
|---|---|
| **Differentiated Management of Quality-of-Service** | 1. Describe the approach used to provide a priority-based differentiated management of quality-of-service? |
| | 2. Describe the approach used to support end-to-end interoperable management of quality-of-service with external networks.<br>Describe any other aspect of the programs QoS support interaction with that of adjacent domains. |
| | 3. What measures of quantitative QoS requirements are supportable, for example jitter, latency, throughput, packet loss, others? |
| | 4. Describe your program's alignment with the DoD QoS/CoS working group roadmap. |

*Table 28. Assess inter-network connectivity*

| | |
|---|---|
| **Inter-Network Connectivity** | 1. What is used as the inter-domain routing Protocol (e.g., BGP4)? How is it being used? Which system interfaces is using it? |
| | 2. Describe the autonomous system boundaries. Describe the interface between internal autonomous systems and external autonomous systems, which are not under your administrative control. |
| | 3. Does the premise router (router that provide the interface between the system being evaluated and external networks) implement features, functions, interfaces, and protocols using open system, non-proprietary methods, if available, that are recognized by appropriate industry standards bodies, working groups, or consortiums? |
| | 4. What proprietary protocols are required on the premise router? |
| | 5. Describe the data interfaces that the premise router supports to external networks. |
| | 6. Describe the methods employed to authenticate routing updates on the premise router. |
| | Describe the filter/access lists to the control plane including route policy updates from external sources. |
| | 7. Describe how the premise router exchanges network reachability information with external networks. What protocols are used? |
| | 8. Describe how the premise router authenticates BGP routing information from external networks. |

In a net-centric development environment, testing procedures should be created that are based upon understanding and maximizing the performance of a service-oriented environment. This allows testing and profiling of a service's many dependencies along with testing each of the web services directly.

Performance testing during a net-centric development process can be carried out in two stages. A quick glimpse of the stages is the following:

- Define metrics that directly measure some element of a web service's performance.
- Create tests that measure individual system components allowing you to determine best methods to increase overall system performance through these defined metrics.

## First Stage Testing

Organizations should start by analyzing the error messages and review differences in error states within web applications and web services. It has been found that web servers reach their breaking point when they are oversaturated with service requests and when they can no longer provide an adequate response. This may cause the server to either go down or the server to simply report that it is unavailable to a large majority of its requests.

Another level of first stage testing is required for service protocols. Testing needs to proceed for SOAP requests as the number of requests are evaluated for a web service failure. Web service errors should also be mapped to HTTP service codes, and the resultant effect on the application environment and the program code.

### Later Stage Testing

The later stage testing consists of testing the actual code residing within application servers. This also includes the runtime environment, the operating system and any middleware that is integrated for messaging and transfer. System testing should result in to modifications to components that can increase overall system performance.

## SOA-Based Testing Framework

In the upcoming future, as SOA-based environments prevail within the industry, there will be the need for an automated services testing framework. This testing framework will provide the proper rigor to fully test large SOA-based infrastructure, as within a military environment.

The idea is to use open standards such as XML and SOAP to define a tool that allows organizations to:

- Perform unit testing for specific modules, to review the fine grained, "atomic" services
- Conduct integration testing the loosely coupled, coarse grained business services that orchestrate the granular services.

Since an agile, dynamic software environment has to revolve around change, development teams need to automatically execute the services' original set of test cases to ensure quality immediately, as peripheral changes are continually being made. To address this need for continual change, a SOA-based testing framework requires the following features:

- Configurable completely via XML
- Use of advanced test input configuration that may need XML input and database state management
- Use of output verification capabilities to verify both for success and for errors
- Publish information based on HTML reports as generated output
- Automated test code generation, unit testing and execution

## Need for Automated Testing

We all know that testing requires a significant amount of time, effort, and discipline. In any software development project, a common set of steps should be followed to assure that testing is occurring properly.

In today's software development environment, we need to take more time in jotting down user requirements than to actually codify these requirements into automated software. Agile development methodologies require that organizations conduct testing on an ongoing basis. This then leads to the need for automated testing. It eliminates the need for error-prone, time-consuming manual tests at the end of development. Automated testing methods should be employed to test a SOA-based environment, especially to assess the functionality of each of the web service middleware layer.

Automated test process should follow a delineated set of procedures and expected results. Best practices should include:

- Code should be integrated nightly and run against suites of automated tests to verify errors on an ongoing basis
- A test plan should be defined that outlines the testing process and exit criteria
- Test cases should be derived from use cases or business requirements
- Test data should be generated from scripts for each test cases
- Expected results for each test case should be outlined
- Test case should be executed and results should be verified to match expected results
- Test reports should measure the software's quality against the test cases
- Defects should be fixed or resolved based on software inadequacies

In a SOA environment, services are highly amenable to testing. Services have well-defined interfaces and well-defined inputs and output. The difference between ordinary software components and services is that services are generally reusable code. This code implements business processes which orchestrate functionality within lower-level components and other finer-grained services. This behavior can use black-box testing methods that are very amenable to automated testing.

## CASE STUDY: UNIVERSITY OF NEW HAMPSHIRE INTER OPERABILITY LABORATORY (IOL)

Within this section we would like to highlight not only the need for assessing and testing service-oriented business processes, but also identify proper locations to conduct this testing. One organization that has become a leader in interoperability testing, in general, is the University of New Hampshire (UNH) Research Computing Center's Inter Operability Laboratory (IOL).

UNH is a state university that established IOL as a part of its Research Computing Center (RCC) in 1988, to specifically address the need for data communications technology research, testing, and education. The IOL has now become a world renowned, highly respected center, which for example, has moved our country's effort forward in IPv6 research and IPv6 adoption.

The following set of questions and answers provide an understanding of the lab and reflect information from the IOL web site, located at www.iol.unh.edu.

- What is the UNH-IOL?

The University of New Hampshire Inter Operability Laboratory (UNH-IOL) tests networking and data communications products. The university established the laboratory in 1988 with the dual mission of providing a neutral environment to foster multi-vendor interoperability and conformance to data communications networking standards while educating students for future employment in the industry. The laboratory has since grown into one of the industry's premier independent proving grounds for new technologies.

- What is a Consortium?

A consortium is a group of companies who come together to test and mutually solve interoperability problems in a specific technology. By working together to provide equipment and to fund development of test methodologies, equipment manufacturers demonstrate their commitment to making more interoperable products thus benefitting their customers and the entire industry while saving each member's finances.

- What is a Testing Service?

A Testing Service, unlike a consortium, provides a contract-based set of tests. Companies wishing to test in a Testing Service pay a flat fee per test suite performed. After a consortium has been around a while and the test suites, test tools, and standards have stabilized to the point where few changes are taking place, the costs and development work associated with the technology are significantly reduced, so it becomes more efficient to operate on a fee-based system where each participant pays only for what they use.

- Where is the UNH-IOL and how do I get there?

Our 32,000+ square foot facility is near the University of New Hampshire campus in Durham, NH, although we operate independently of the university's academic departments. . The IOL is close to airports in Boston, MA, Manchester NH, and Portland, ME. It is also accessible via train and bus. Directions to our facility area are available here.

- Who works at the UNH-IOL?

More than 100 graduate and undergraduate student-employees from the University of New Hampshire work with full-time UNH-IOL staff, gaining hands-on experience with developing technologies and products from hundreds of major companies.

- Industry Associations

The University of New Hampshire Inter Operability Laboratory is involved in many industry associations related to the technologies we support. Our activity in these associations keeps us on the cutting edge of the latest developments in technology. We have also been able to contribute to these associations through our knowledge and experience. This participation demonstrates our commitment to the advancement of technology, and enhances our testing ability.

- Standards Involvement

As active members of many standards organizations, the UNH-IOL has been able to effect positive change to several standards. Many of our technology experts have served as editors for new and established standards. This participation keeps our staff well informed in all aspects of the standards that we test.

## REVIEW OF CHAPTER GOALS

The goals of this chapter were to address:

1. How we assess compliance to net-centric data strategy:
   - Make Data Visible, Make Data Accessible, Make Data Understandable, Make Data Trustable, Make Data Interoperable, Provide Data Management, Be Responsive to User Needs, Obtain Reliable and Secure Information
2. How we assess compliance to net-centric services strategy:
   - Address service-oriented architecture, meet need for open architecture, address scalability and availability, accommodate heterogeneity, perform decentralized operations and management and enterprise service management
3. How we abide by net-centric information assurance assessment:
   - Net Centric IA posture and continuity of operations, identity management, authentication and privileges, privilege management, mediate security assertions, cross security domains exchange
4. How we address communications and transport assessment:
   - Figure out IPv6 compliance, packet switched infrastructure, layering and modularity, transport goals, network connectivity, concurrent transport of information flows, differentiated management of quality of service, inter-network connectivity
5. How we evaluate interoperability testing for net-centric development:
   - Ensure first stage and second stage testing, enable automated testing, and SOA-based testing

## REFERENCES


Bridges, S., Zeigler, B. P., et al. (2005). *Enterprise Infrastructure for Model & Simulation-Based Testing of Net-Centric Systems*. White Paper, Joint Interoperability Test Command, Ft. Huachuca, AZ.

Eleazer, B., & Reeder, S., (2007, April). *Testing Concept of Operations (CONOPS) in DoD's Net Centric Environment Testing*. 8thAnnual NDIA Science & Engineering / DoD Tech Expo.

Office of the Assistant Secretary of Defense for Networks and Information Integration, Department of Defense Chief Information Officer, (2003). *DoD Net-Centric Data Strategy* (Version 1.0).

Office of the Assistant Secretary of Defense for Networks and Information Integration, Department of Defense Chief Information Officer Directive 8320.02-G, (2006, April). *Guidance for Implementing Net-Centric Data Sharing.*

Office of the Assistant Secretary of Defense for Networks and Information Integration. Department of Defense Chief Information Officer, (2006, April). *Clinger-Cohen Act Compliance Certification of Major Automated Information System for Fiscal Year2006.*

Office of the Assistant Secretary of Defense for Networks and Information Integration, Department of Defense Chief Information Officer Directive 8320.02-P, (2007, April). *Data Sharing in a Net-Centric Department of Defense.*

Office of the Assistant Secretary of Defense for Networks and Information Integration, Department of Defense Chief Information Officer, (2008, March). *The Department of Defense Interim Information Assurance Strategic Plan.*

Office of the Assistant Secretary of Defense for Networks and Information Integration. Department of Defense Chief Information Officer, (2008). *Department of Defense Information Management/ Information Technology Strategic Plan 2008-2009.*

Office of the Assistant Secretary of Defense for Networks and Information Integration, Department of Defense Chief Information Officer (2007, July). *Net-Centric Checklist* (Version 2.1.4).

The MITRE Corporation. (2005, July). *NCOIC White Paper*. Moser, S., Lt. Col., (2005, November) *Interoperability Certification Process and NR-KPP Brief*, DoD NII. Gaetjen, T., Lt. Col. (2005, May). Army G6, *Net-Ready Key Performance Parameter*, DoD Briefing Presentation.

# Chapter 16
# Technology Evolution Assessment for the Future

## CHAPTER CONTENT

As you explore Chapter 16, it will cover the following topics:

- The Technology Evolution Process
- Assessing Key Technology Areas
- Net-Centric System Technology Forecast
- Acquisition Trade Study Process
- CASE STUDY: Net Centric Operations Industry Consortium (NCOIC)

## CHAPTER FOCUS

This next chapter addresses assessment describes the technology evolution process that involves government and corporate firms to perform technology validation and evolution planning. We go ahead and assess key technology areas, and provide a system technology forecast that can be used by a repre-

DOI: 10.4018/978-1-60566-854-3.ch016

sentative organization on the road to net-centric readiness. We then mention the acquisition trade study process and provide an understanding as to how vendor products can be assessed in an objective and documented manner. We end the chapter by providing a case study and information on the Net-Centric Operations Industry Consortium (NCOIC).

## THE TECHNOLOGY EVOLUTION PROCESS

Technology evolution is an obvious and yet a widely encompassing concept. In the case of a net-centric program or system, it is understood that the system has to stand the test of time. This is because information in net-centric systems within the government, military and corporate world has to be kept ongoing for a long period of time. Most government agencies have archival and storage retention rules for official records. For example, the Department of Veteran's Administration has a "default" 75-year retention rule from the last date of activity, with regard to archival content. This stems from the fact that military veterans enlisted with the VA needs to be has their records stored for this period of 75 years. Retention rules vary for different federal agencies and state governments, however, at least a 25-year retention rule applies for a significant amount of content.

Based on these long retention rules, it is clear that digital technology used within net-centric systems will have a much faster change and evolution process, than the system maintenance and upkeep period. In this case, government and industry personnel need to address technology evolution as a primary task in their governance of net-centric systems.

A dedicated technology evolution process needs to be adopted for any net-centric system to ensure that digital content stored within an archive can be accessed at any time. Within a large, dedicated hardware and software infrastructure, there needs to be a continuous planning process to ensure that the latest tools are being used to stay current with evolving technology.

### Government and Corporate Evolution Planning Process

This section address the evolution planning process recommended for government and corporate organizations. The idea is to address methods used to develop research and commercial marketplace information and assessment data relevant to the development of the evolution plan. As part of the incremental development life cycle, it is important for government and corporate program teams to conduct periodic reviews with key technology vendors, researchers, and to review new technologies, research, and product line development roadmaps and plans.

On an ongoing basis, an Integrated Product Team (IPT) can be formed that has representatives from government personnel, vendor management, consultants, engineers and business and user community. The role of the IPT is then to assess the stability of commercial product suppliers and the status of particular products in the system design to make sure these products continue to be available. These assessments include reports on the impacts of changes to widely used commercial standards.

The technology and standards information that is collected is assessed based on their maturity, cost of ownership, and projected usefulness to the net-centric system. The value of newly introduced products and research activities is balanced with the stability of the technology. In many cases, the first few years of a technology are exposed to a great deal of instability that may not be suitable for the low risk-profile that most programs may to want to establish for development and sustainment.

*Figure 1. Technology Validation and Evolution Planning*

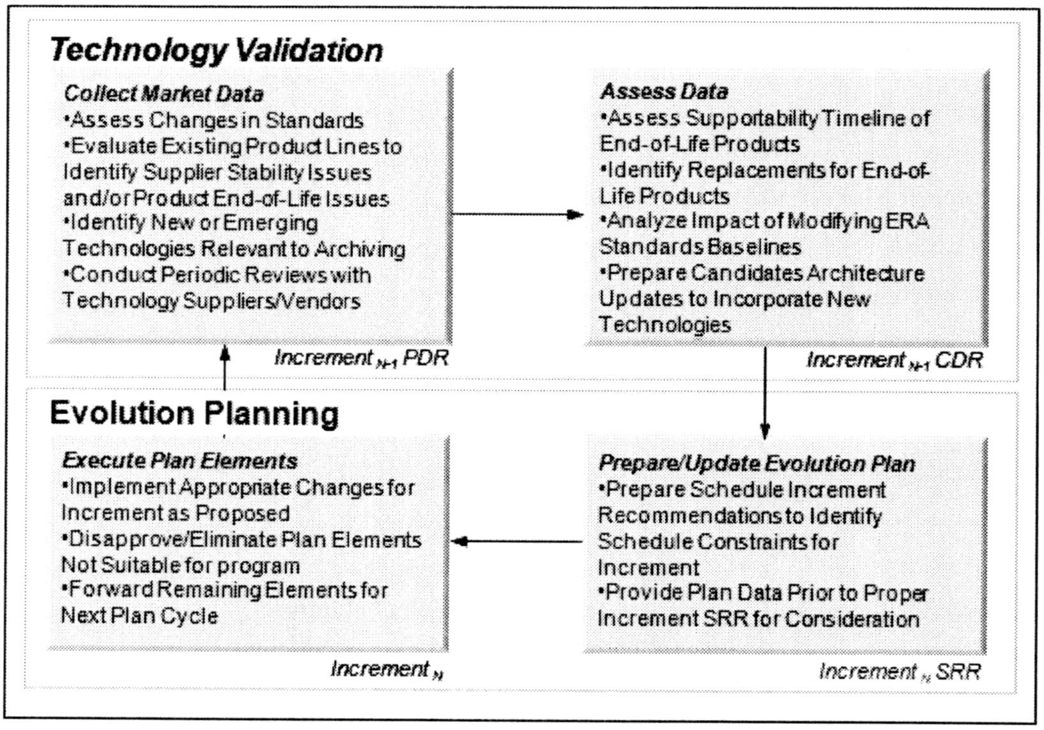


The technology evolution effort is focused most heavily on those aspects of the system where new requirements are emerging or current technology is facing obsolescence.

These processes need to be fully integrated within the system design and development life cycle. During most large development projects, it is recommended that there be a set of design reviews. These should be Preliminary Design Reviews (PDRs), and Critical Design Reviews (CDRs) that may be conducted for each increment of a program. System requirements also require reviews, for example in a System Requirements Review (SRR).

Figure 1 below illustrates technology evolution planning as part of the requirements review and design review for a large system developed in increments. It addresses two aspects of the process, technology validation and evolution planning. The expectation is that the hardware and software resident within the system undergoes a technology review process during the design reviews held for the system. This then helps in defining the new set of hardware and software that is necessary for the next increment.

There is a definite link between the technology validation process and the evolution planning process. As shown in the diagram, the following steps should be followed for technology validation:

- **Collect Market Data**
  - Assess Changes in Standards
  - Evaluate Existing Product Lines to Identify Supplier Stability Issues and/or Product End-of-Life Issues
  - Identify New or Emerging Technologies Relevant to Archiving
  - Conduct Periodic Reviews with Technology Suppliers/Vendors

- **Assess Data**
  - Assess Supportability Timeline of End-of-Life Products
  - Identify Replacements for End-of-Life Products
  - Analyze Impact of Modifying System Standard Baselines
  - Prepare Candidates Architecture Updates to Incorporate New Technologies

The evolution planning process may be implemented by a separate group such as the engineering team who is fully apprised of the hardware and software tools in use by the enterprise. The evolution planning activities include the following steps:

- **Execute Plan Elements**
  - Implement Appropriate Changes for Increment as Proposed
  - Disapprove/Eliminate Plan Elements Not Suitable for program
  - Forward Remaining Elements for Next Plan Cycle
- **Prepare/Update Evolution Plan**
  - Prepare Schedule Increment Recommendations to Identify Schedule Constraints for Increment
  - Provide Plan Data Prior to Proper Increment SRR for Consideration

In working together, the expectation is that the correct set of technology is devised and used for a system as we march forward in time. This also aids in technology refreshment during the sustainment of the system as it is deployed and integrated within the existing infrastructure.

## ASSESSING KEY TECHNOLOGY AREAS

This section provides an inkling moving forward as to the types of technologies that are important to evaluate as part of the overall net-centric transformation. These technologies, however, have wide use across multiple industries in commercial organizations.

An assessment of these technologies can be made based on overall needs of the DoD and military, along with government contractors that are aiding in this process. The military to civilian technology transfer has been active in fomenting these activities within the respective industries.

- **Search Technology** – This topic addresses the need to conduct enterprise-level searches throughout an organization. It concentrates on information retrieval that focuses on developing frameworks to represent documents, queries, and their relationships. Search technology is moving to incorporate multimedia content such as images, audio and video and the ability to easily search based on a combination of media types. A search system needs to provide the ability to correctly identify and score relevant documents with respect to a given query based on the user's primary needs. Relevancy is a function of recall and precision, and an effective search system needs to balance these with activities such as proximity searches, concept searches, and the use of a dynamic taxonomy structure for the searchable content.
- **Enterprise Content Management (ECM)** – ECM is a composite technology concept that allows organizations to capture, manage, store, preserve, and deliver content and documents

related to business processes. The use of ECM is to address records management and the archival life cycle to ingest, store, disseminate and retrieve digital assets. The use of a loosely-coupled SOA-based interface is important and needs to be integrated within the ECM application and maintenance. The consequences of software middleware activities such as use of workflow, orchestrations, controls and data integration need to be fully understood for the upcoming future.

- **Preservation File Formats** – this topic may not be a defined technology area, but it is of primary importance to digital archiving and preservation. The use of open standards is preferred, and as digital document formats evolve, it is important to understand the level of metadata that is necessary to retrieve and search the document. Text document formats using XML, and postscript document formats using PDF are the current state. Use of Office applications in formats such as ODF and OpenXML is the upcoming activity. The best way to preserve current electronic documents is still being debated and there is no current consensus within the industry.
- **Web Services** – This has been a known buzzword in the industry for a while and technology concepts of SOA, object-oriented programming and interfacing, and code publish and subscribe activities, have matured. There are a number of WS standards that prescribe open methods for performing interoperability, security, architecture, choreography and others and these are all being discussed within industry working groups. The evolution of core standards such as SOAP, WSDL, and UDDI need to be tracked. Interface control based on Web Services integration with external entities needs to be specified as technology progresses.
- **Data Security** – In this case data security includes data encryption and what happens to data at rest and data in motion. For federal agencies other than DoD, NIST standards publications and policies need to be reviewed. Use of transfer mechanisms such as SFTP and SCP needs to be tracked for file transfer. The consequence for federal mandates such as HSPD-12 needs to be understood. The rollout of a full PKI infrastructure to protect hardware and software based on the use of common access cards also needs to be tracked.
- **Data Management** – This includes the use of a database management system such as Oracle and the use of a flat-file system that does not use a relational model. The advent of object-relational concepts for data integration and the use of XML documents, complete with a XML Schema specification needs to be explored. Data management includes the ability to associate the proper metadata with digital assets and the ability to conduct system operations such as queuing services, directory services, and addressing localized data stores.
- **Server Processor Technology** – This is to address server hardware platforms and explore where server technologies are proceeding in terms of computer processor power, performance, and scalability. Current servers are changing in processor type, processor speed, use of different operating systems, input/output bus rates and associated technology such as server virtualization, clustering, self-healing attributes, and storage integration.
- **Disk Storage** – There is a need for tracking hard disk storage and its usage for long-term archival along with short term system operations. Disk technology requires redundancy based on RAID standards and its path to RAID-5 and RAID-6. It is important to track inexpensive SATA drives and understand the progression to SATA II. Use of Fibrechannel connections and the connection to storage networks in a SAN or a NAS configuration is also important. The use of storage resource manager software allows storage life cycle management for near-line, and off-line to tape configurations.

- **Tape Storage** – Current tape technology recommends use of a standards based approach and two of the leading tape based removable media formats are LTO and DLT. Tape vendors currently have roadmaps for tape density expansion and the LTO progression moves from LTO3 to LTO 4 to LTO 5. Along tape storage media, tape library consoles are also evolving that allows expansion of tape media as more digital assets are stored. The tape storage has to be managed by storage manager software that determines the optimal methods to store information on an ongoing basis.
- **Network Bandwidth and Protocol** – Technology regarding network communications is a very large topic, so this area focuses on two activities that are of primary importance. This includes the growth of LAN and WAN bandwidth to Gigabit Ethernet, Fiber-based backbones and other communications bandwidth expansions. It also addresses network protocol and specifically the tracking of TCP/IP and its current IPv4 technology. The expectation is that IPv4 will transition to IPv6 standards and all government and private networks will be able to work both with IPv4 and IPv6-based routers. Native IPv6 communication is currently in the long term since it will take a concerted industry approach to use all of the capabilities that IPv6 offers.

## NET-CENTRIC SYSTEM TECHNOLOGY FORECAST

Based on the technology evolution process and the key technology areas mentioned, we can now provide a completed System Technology Forecast as an example within this book. It defines the underlying current and expected supporting technologies. The objective is to provide an understanding of expected supporting technologies that can be reasonably forecast given the current state of technology and expected improvements.

This System Technology Forecast addresses the key technologies identified above and provides a brief set of predictions about the availability of emerging technological capabilities and about industry trends in specific time periods. These time periods are in Short-Term, Mid-Term and Long-Term periods. This forecast uses a 2 year time period that breaks it down as 2006 to 2008, 2008 to 2010, and 2010 to 2012. The ERA System Technology Forecast works with the technology classification necessary to support the program and the technical standards profile identified for the ERA program.

A technology forecast such as the one provided by Table 1 allows organizations to form a consensus opinion by having this information be reviewed by subject matter experts and vendors in the field who provide products within each of these technology areas. The idea is to aggregate this as a composite view of the predictions for the short term, mid-term and long term forecast.

## ACQUISITION TRADE STUDY PROCESS

This section addresses technology evolution as it relates to the acquisition process for buying or building systems. As part of net-centric assessment, it is important to make sure that systems that are built or acquired meet a full set of criteria for a trade study evaluation.

When a large system is either built or purchased there are a number of key architectural decisions that lead to a "buy" rather than a "make" action. The idea is for your organization to objectively consider a number of criteria before making any recommendation. This methodology goes beyond

*Table 1. Net-Centric System Technology Forecast*

| Net-Centric System Technology Forecast | | | |
|---|---|---|---|
| **Technology Area** | **Short Term (2008 to 2010)** | **Mid Term (2010 to 2012)** | **Long Term (2012 to 2014)** |
| **Search Technology** | metadata-based full-text search | dynamic taxonomies | semantic search, audio content search, image content search |
| **Enterprise Content Management** | Common set of content management tools | Integrated ECM, workflow and orchestration | Archival and records management components within ECM |
| **Preservation File Formats** | XML, PDF | XML, PDF/A, ODF, OpenXML | XML, PDF/A, ODF, OpenXML |
| **Web Services** | Registry-based discovery | Enterprise Services; WS-Interoperability | WS-* pervasive adoption |
| **Data Security** | FIPS 140-2, AES | FIPS 200, HSPD-12 | HSPD-12 based PKI |
| **Data Management** | relational database, object-relational database | object-relational database | native XML database |
| **Server Processor Technology** | multi-core CPU | server virtualization, blade servers | commodity grid computing |
| **Disk Storage** | SATA, FibreChannel, 500 GB disks | SATA II, RAID-6, 750 GB disks | RAID-6, 1 TB disks |
| **Tape Storage** | LTO 4 | LTO 5 | LTO 6 |
| **Network Bandwidth and Protocol** | Gigabit Ethernet, IPv4 | 10 Gig Ethernet, IPv4 to IPv6 transition | Upcoming Broadband IPv6 activities |

the ordinary feature/function comparisons to include supplier commitment and stability, market acceptance, product maturity, and compatibility with your system's architectural principles.

As part of the assessment, the idea is to ensure that services and solutions that are acquired follow a structured mechanism. Table 2 lists below a set of architecture evaluation criteria to properly conduct a vendor trade study.

During the evaluation process, these criteria provide the basis for assessment and ranking so that the highest ranked criteria exert the most influence on the evaluation. Actual Commercial-Off-The-Shelf (COTS) hardware and software recommendations are part of acquisition and buying decision based on a formal trade study process.

Figure 2 provides a flowchart of the trade study process as can be conducted in large corporate or government organizations. A trade analyst refers to staff personnel within the organization that is part of the actual decision making.

Here is a stepwise process flow:

- **Define Problem** – The Trade Analyst defines and documents the purpose, objective(s), constraints, and requirements of the trade study, including definition of the need, the user, and the availability of resources bounding the scope of the analysis. Without a clear statement, studies become costly and produce unclear results.
- **Establish Evaluation Criteria** – The Trade Analyst establishes, documents, and maintains the criteria for evaluating alternatives, and the relative ranking of these criteria.
- **Identify Alternative Solutions** – The Trade Analyst identifies and documents alternative

*Table 2. Trade Study Architecture Evaluation Criteria*

| Criterion | Description |
|---|---|
| Architectural Compliance | Compromises to the architecture reduce flexibility of the design and introduce risk to the program. Candidate products for the component must adhere to the service-oriented architectural principles and comply with the selected key technologies and standards. Candidate products must demonstrate ability to meet performance requirements. Where applicable, the NARA Technical Reference Model component is identified for the candidate product. |
| Functionality | Products must provide functionality as allocated from the SyRS requirements and the Service Methods defined in Section 4.2. Products must demonstrate the ability to meet performance requirements. |
| Openness, Scalability, Supportability | Candidate products must adhere to open systems-based standards and have already established their ability to scale to NARA's requirements. The supplier must demonstrate an ability to support its product in the government sector. |
| Portability | The candidate product must be available for multiple computing platforms so that ERA has the freedom to alter hardware and operating system software without being forced to alter COTS applications. |
| Technical and Administrative Supportability | The candidate vendor must be able to provide administrative support, which includes help desk materials, course materials, and customized training, where applicable. Technical support includes support during hardware and software installations, technical upgrades, and maintenance activities. |
| Level of Market Acceptance | The market must have demonstrated acceptance of the supplier and candidate product to at least the degree that the supplier has an installed base that lends credence to the product's use in the ERA System. |
| Supplier Stability | The product supplier must be financially sound and be able to maintain a commitment to the candidate product. |
| Risk Mitigation | The LM Team may recommend a COTS product that does not meet all of the criteria if the alternative is costly or requires time-consuming development. |
| Cost | Applies the engineering knowledge base from the life cycle cost model to the established cost objective to manage risk and reduce ownership costs. |

solutions to the problem.

- **Define/Select Evaluation Methods** – The Trade Analyst selects, defines, and documents the methods for evaluating the alternative solutions against the established criteria. The Trade Analyst also ensures that the selected evaluation methods are commensurate with the resources needed for the scope of the analysis.
- **Define Criteria Weights** – The Trade Analyst defines and documents weighting factors for the evaluation criteria
- **Evaluate Alternative Solutions** – The Trade Analyst evaluates alternative solutions using the established criteria and weighting, and documents these evaluations.
- **Sensitivity Analysis** – The Trade Analyst determines the sensitivity of the solution ratings to small changes in input values.
- **Determine Study Completion** – The Trade Analyst determines study completion by ascertaining whether the analysis is complete and sufficient to permit solution selection or whether further analysis is needed.
- **Select Solution(s)** – The Trade Analyst selects a solution(s) from the alternatives based on the evaluation criteria, and documents the rationale for this selection.

## CASE STUDY: NET-CENTRIC OPERATIONS INDUSTRY CONSORTIUM (NCOIC)

A number of government and industry forums have been formed to address the issue of Net Centricity and its proliferation within the government and civilian marketplace. This case study examines an organization that was created specifically to tackle this issue, and it is the Net-Centric Operations Industry

*Figure 2. Trade Study Process for Vendor Purchasing*

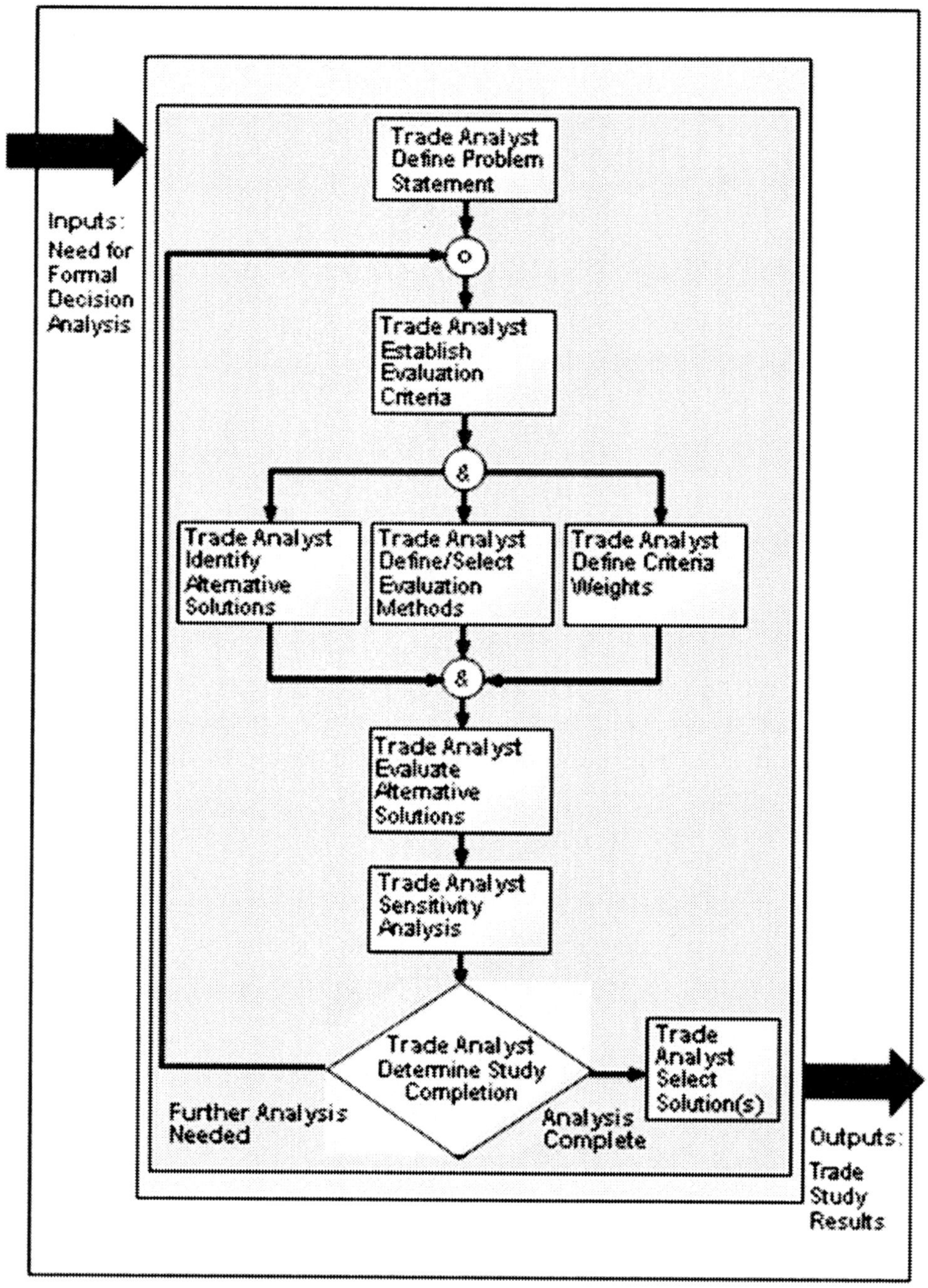

Consortium (NCOIC). Here is a set of information that has been paraphrased from their web site that details the background, mission and goals of the organization.

## Background

The Network Centric Operations Industry Consortium (NCOIC) is a not-for-profit international entity that is committed to convening members of industry to integrate existing and emerging open standards into a common evolving global framework that employs a common set of principles and processes to assist with the rapid global deployment of network centric applications. Established in 2004, the NCOIC consists of industry representatives from defense companies, large-scale systems integrators, information technology providers, and academia working in concert with advisory bodies consisting of government officials, standards groups and other stakeholders.

NCOIC information and consortium activities can be found at www.ncoic.org

The Network Centric Operations Industry Consortium, (NCOIC) is an international, industry-wide consortium that combines the experience and expertise of a large number of major companies. It recommends a unified approach that may enable sensors, and allow communications and information systems to interact within a global network centric environment.

## Why Have a NCOIC?

The mission of NCOIC is to bring together the best of industry in a collaborative forum. NCOIC intends to recommend and provide input on the underlying standards and architectural approach that systems and platform developers may follow to enable each platform, system or application to participate in a global network environment.

Through the NCOIC organization, the members intend to jointly identify open standards, their patterns of use and interoperation, and common practices, processes, and principles for aiding the DoD and the military in net-centric force transformation. Creating a standards-based information backplane increases competition in the vendor community. Instead of segmenting the market, and closing it to whole groups of competitors, this effort will level the playing field so that all can create products to differentiate themselves. These products will be focused on domain solutions, as they will not be recreating the system-level infrastructure each time. NCOIC complements and enhances the efforts of other, ongoing industry and government associations and initiatives behind force transformation and homeland security.

## NCOIC Tasking Strategy

NCOIC has tasked itself to demonstrate how the net-centric industry can converge through the employment of interoperability of systems based to the greatest extent possible on the use of well-defined open standards. NCOIC has proposed tasking its member organizations based on four initial parallel strategies. These activities have already been vetted by members within both industry and government:

1. **NCOIC Strategy 1** – Complete a thorough and rigorous analysis of pertinent government agency architectures, requirements, and mandated open standards to identify commonalities, conflicts, gaps, and areas for potential improvement.

2. **NCOIC Strategy 2** – Complete a broadly inclusive information management overarching architectural framework and continuously identify more open standards, to enable ever increasing degrees of NCO, as rapidly as possible, now and in the future. The overarching architectural framework and open standards must support re-usable solution models that can evolve in the future, and be scaled and/or replicated, rapidly and cost-effectively, for every enterprise.
3. **NCOIC Strategy 3** – Use the open standards-based architectural framework to facilitate acceleration of the accomplishment of NCO by the implementation of Commercial Off-The-Shelf (COTS) or Government Off-The-Shelf (GOTS) and other open source products whenever possible, and to encourage the future development of additional open source items in the future.
4. **NCOIC Strategy 4** – Develop a program for education and outreach to continuously increase the awareness, adoption, and use of the open standards identified by NCOIC, and the necessity to change traditional thinking, approaches, and cultures, to further accelerate NCO.

NCOIC intends to create value for all its pertinent stakeholders:

- Global government customers
- NCOIC membership of large-scale system integrators, platform providers, and product and service companies
- Citizens

Here are a set of domains that provide an understanding of how NCOIC plans to provide thought leadership to net-centric operations (NCO).

## The Knowledge Domain

NCOIC will build knowledge of NCO, thereby facilitating an understanding of existing and emerging capabilities by:

- Incorporating the "boundary-less" concept into NCOIC logic
- Building understanding of the full NCO landscape – technical and strategic
- Building awareness of how NCOIC can enable all aspects of NCO
- Creating outreach and educational initiatives to educate the community about NCO technologies services, and solutions

## The Enablement Domain

NCOIC will work to enable network-centric compliant products, services, and processes by fostering innovative NCO concepts and strategies to meet mission challenges. NCOIC will lay the foundation for rapid, agile, and nimble product development aimed at enabling NCO at all levels according to the value stream of each NCOIC member, individually and/or combined.

Enabling the creation of network-centric compliant products will provide a stream of compatible technologies for procurement. Transformation to an integrated operating environment will then become a natural progression following the knowledge and enablement tenets.

### The Transformation Domain

NCOIC will be active in promoting the adoption and implementation of NCOIC concepts. Its active role will assist in transforming current operating environments to NCO via the implementation of NCO concepts and strategies into operational capabilities. NCOIC will:

- Expedite the development, testing, and refinement of concepts and strategies into solutions that are enablers for transformation,
- Work in close partnership with stakeholders and customers to focus on the development and deployment of mission-capable systems into the NCO architectures,
- Work with customers to help revise the acquisition processes and transform other institutional barriers.

## REVIEW OF CHAPTER GOALS

The goals of this chapter were to address:

1. What does a technology evolution process signify?
   - Technology has to withstand the test of time as it evolves and hardware and software solutions change on a continual basis
2. How have government and large corporate firms addressed the technology evolution planning process?
   - Understand the technology validation and evolution planning process as it is incorporated within the systems development lifecycle
3. What are some of the key technology areas that have been identified moving forward for technology planning purposes?
   - Address search technology, enterprise content management, preservation file formats, web services, data security, data management, server processor technology, disk storage, tape storage, and network bandwidth and protocol
4. What is a system technology forecast with regard to net-centric readiness?
   - A system technology forecast addresses short term, mid-term and long term technology industry trends
5. What is involved in an acquisition trade study process?
   - Address how a deliberate trade study process allows organizations to lead to a make vs. buy decision
6. Understand the acquisition trade study process that leads to a purchasing decision from major vendors.
   - Evaluate the stepwise process flow that should take place to identify and evaluate product solutions

## REFERENCES

America, I. T. S. (2005, October). *Technology Evolution: Lessons Learned and their Impact on VII Applications*. Retrieved from http://www.itsa.org/itsa

Ghosh, S. (2006, July). *Electronic Records Archive (ERA) Technology Evolution Plan*. National Archives and Records Administration, Lockheed Martin, 2nd ERA Technology Evolution Council Meeting.

Management, I. & Policy Development Workgroup, Health Information Exchange (2006, September). *Task Force to Study Electronic Records*. EMC White Paper (2008, August). *Enterprise Information Management: Information Virtualization for a Unified Business View*. Department of Defense (2003, October). *Joint Technical Architecture,* (Vol.1). (Version 6.0). Faughn, A.W., (2002, October). *Interoperability: Is it Achievable?* Program on Information Research Policy, Harvard University.

The MITRE Corporation. (2005, July). *NCOIC White Paper*.

# Chapter 17
# Achieving a Net-Centric Service-Oriented Enterprise

## CHAPTER CONTENT

As you explore Chapter 17, it will cover the following topics:

- Defining a Net-Centric Service-Oriented Enterprise
- Transition Mechanism based on Life Cycle Processes
- Service Oriented Enterprise Technology Features
- Example Set of Net-Centric Services
- Changes and Benefits to Different Industry Sectors

## CHAPTER FOCUS

This last chapter focuses on the target state of achieving a Net-Centric Service-Oriented Enterprise (NCSOE). All of our previous discussions led to the point of defining this target state, the transition processes that allow us to get there, and the benefits of the future state. The transition mechanism

DOI: 10.4018/978-1-60566-854-3.ch017

provides a life cycle methodology that includes a business roadmap, service enablement, and service sustainment. We describe each of these transition processes, their inputs and outputs. We then discuss a number of the technology features that we obtain based on a service oriented enterprise. We provide example net-centric services that can be defined as the common set for the overall enterprise. We then provide a set of benefits of NCSOE to different industry sectors.

## DEFINING A NET-CENTRIC SERVICE-ORIENTED ENTERPRISE

So far as a reader, you have read about the promise of net-centric transformation, but the target state has been hazy, since a specific description has not been given as to its formulation. Well, this chapter delineates this target state as the Net-Centric Service Oriented Enterprise (NCSOE). This is a new generation of the information enterprise that is now making headway within the commercial marketplace. Within the upcoming years, the market will further define the characteristics of an NCSOE.

At this time, the industry intention is to capitalize on the new technology changes brought forth by paradigm shifts such as the move from IPv4 to the IPv6-based Next Generation Internet. This allows large organizations to raise their enterprise operational efficiency to a whole new level. Our methods of information sharing and collaboration are also being pushed to a new level. It is no longer necessary for each of us to be working together real-time in offices and factories since we can choose to share information in an asynchronous manner at our own time. Decision making can rapidly take place at remote centers through the use of an expanded set of collaboration activities and the informed consent of management and leadership. The target NCSOE state allows us to make these dreams come true.

### Defining the Target State

Net-Centric Service Oriented Enterprise takes advantage of the current DoD direction to create a Global Information Grid and capitalize on broadband communications and capabilities. It allows us to change our performance outcomes that lead to organizational adaptation, flexibility and decision making. NC-SOE allows organizations to improve their ability to compete in the marketplace, or for the military to rapidly formulate assaults on real-time threats in the battlefield. NCSOE allows us to not ignore the technological trends of converging business and process management so that we not jeopardize our competitive edge.

NCSOE provides focus and direction to current enterprise technologies such as distributed computing and the transition from Object-Oriented Architecture (OOA) to Service-Oriented Architecture (SOA). Within the commercial and consumer market, there has been a rapid increase in information exchange across wired and wireless networks. This also raises the need for a greater infrastructure of broadband, high-demand networks that provide information in a timely and safe manner. The purpose of NCSOE target state is to provide the following:

- Provide an enterprise-level information collaboration solution that can enforce decision priorities and quality of service
- Provide a decentralized, loosely coupled enterprise that is highly interoperable
- Provide a system-of-systems environment that has a synergistic combination of data and system interfaces to allow rapid information processing

*Figure 1. Transformation to the Target State*

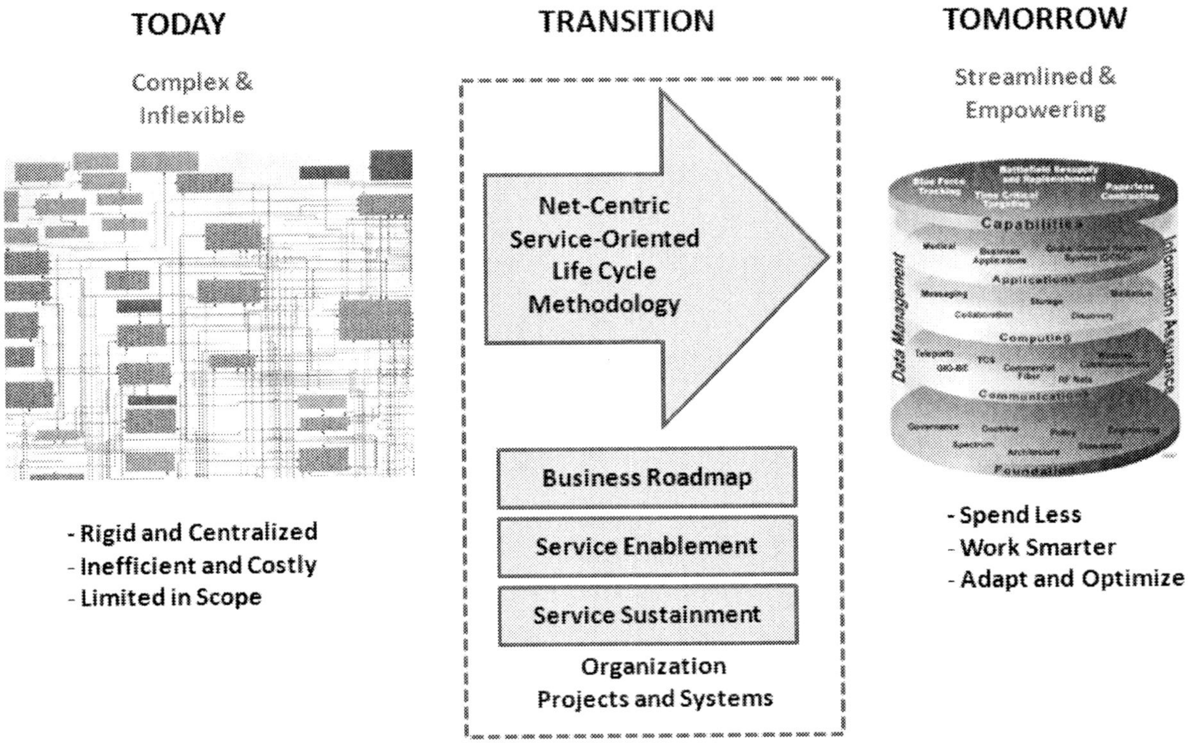


## Formulating the Transition Mechanism

To transform our current enterprise to the NCSOE target state, we need to prepare a worthy transition mechanism. It is best to focus on a business-centric approach where technology is adapted to support business stakeholders, customers and end users. The intent is clearly to move toward a more flexible architecture that is less costly to maintain.

Figure 1provides a depiction of the current state of the enterprise today that is complex and inflexible, a formulation of our transition mechanism and the promise of tomorrow's enterprise as being streamlined and empowering.

As the diagram illustrates, the systems and communications networks of today are overly complex, that cause information flow to be rigid and within their own stovepipes. This also limits the scope of the data that is resident within each stovepipe since a broad set of users cannot make use of the existing information.

The transition mechanism uses a net-centric service-oriented life cycle methodology to move the enterprise forward for a large or complex organization. The idea is to prepare a business roadmap, enable the use of enterprise services and be able to maintain and disseminate these services with other organizations or enterprises.

The target state uses the net-centric data strategy depiction, where enterprise capabilities are offered in layers. We start with the foundation layer that incorporates all of the business rules of the enterprise, a communications layer that provides a choice of different network topologies, a computing layer that creates middleware that orients itself toward use of enterprise services, the application layer that prepares all of the business and consumer applications, and the capabilities layer that provides you with ongoing decision making abilities. The implementation of these abilities within a NCSOE state allows you to spend less, work smarter and optimize your solutions quicker to meet the needs of the marketplace.

## TRANSITION MECHANISM BASED ON LIFE CYCLE PROCESSES

This section details a set of processes and activities that need to be performed to transition to a service-oriented enterprise, and the number of strategic activities that encompass these processes. It is expected that the business processes can serve as a roadmap for major organizations that have a complex enterprise and manage a portfolio of system capabilities and investments.

The life cycle methodology provides us with the transition mechanism to prepare ourselves for the NCSOE target state. It consists of three major components:

- **A Business Roadmap** – Provides overall direction for the organization to develop with regard to a service-oriented enterprise. It develops business objectives, creates a service-oriented vision, and prioritizes business projects and tasks to overhaul the current enterprise.
- **Service Enablement** – This activity provides the processes for SOA delivery. The enterprise has to change in its existing business processes to realize the full benefits of this shift in technology implementation.
- **Service Sustainment** – Organizations currently sustain their information systems based on maintenance plans and the ongoing needs of their users. This service sustainment allows the organization to maintain a SOA, with all of the support processes and infrastructure. As a SOA environment matures, service sustainment also provides the path to robust adoption and technology evolution strategy.

Each of these life cycle processes will be further detailed within this section.

### Description of Roadmap Activities

Figure 2 provides a business roadmap that focuses on providing overall direction to developing and maintaining a set of services within the enterprise.

It is denoted into four main groups of strategic activities and processes: Federated Vision & Strategy, Program Management, Federated Enterprise Architecture, and Executive/Portfolio Management.

### Federated Vision and Strategy

A federated vision allows each sub-organization within a large enterprise to work in an autonomous manner. Using a SOA strategy, federation ensures that decision makers and program managers are able to align their programs and capabilities across the various tiers of the overall enterprise. Table 1 outlines the processes and the inputs and outputs to and from each process.

*Figure 2. Service-Oriented Business Roadmap*

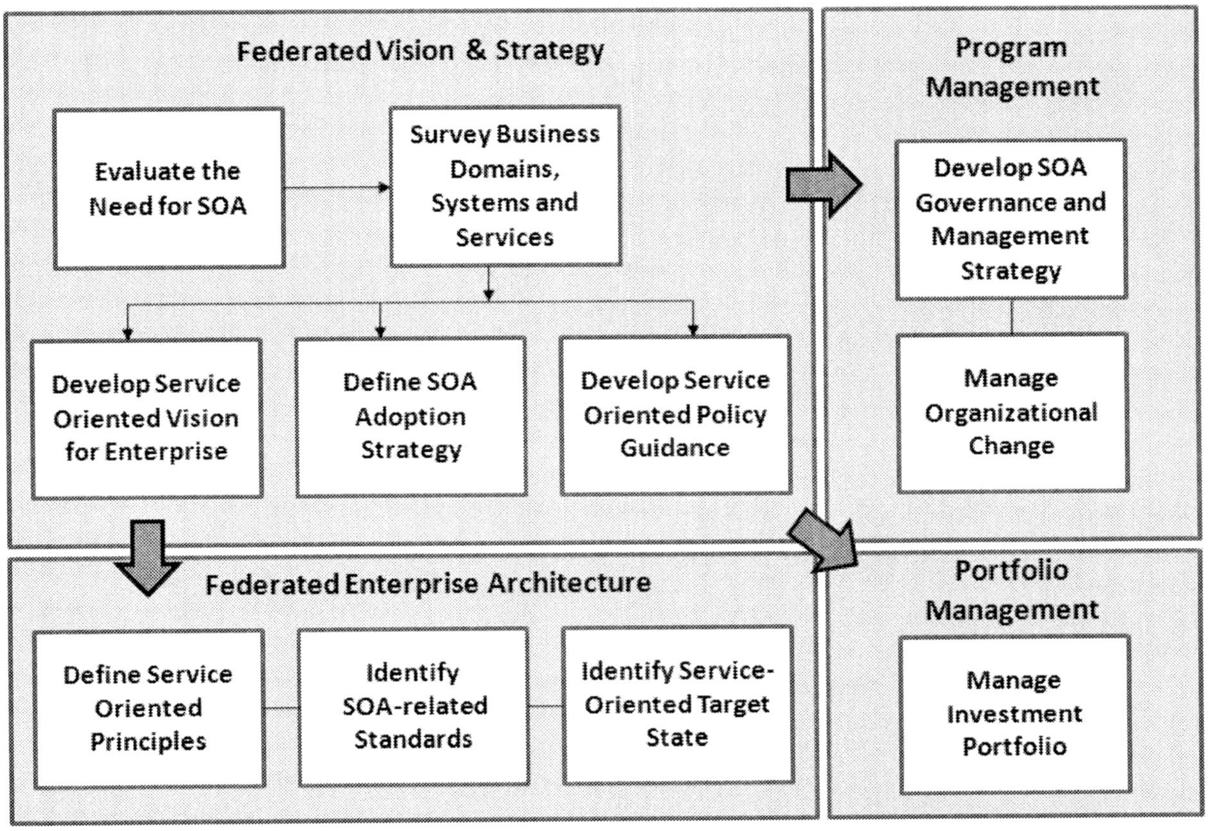


*Table 1. Federated Vision and Strategy Processes, Inputs and Outputs*

| Processes | Inputs | Outputs |
|---|---|---|
| Evaluate the Need for SOA | - High-level business needs<br>- Business issues<br>- Strategy and vision documents<br>- Business capability descriptions<br>- Enterprise architecture | - Completed needs analysis of service capability<br>- SOA component characteristics to fulfill needs analysis |
| Survey Business Domains, Systems and Services | - Documentation of current business domains, systems and services<br>- Business needs<br>- Community of Interest (COI) expectations<br>- Service-oriented vision | - Asset inventory<br>- Business category descriptions such as scenarios, case studies |
| Develop Service-Oriented Vision for Enterprise | - SOA needs analysis<br>- Business capability descriptions<br>- Asset inventory | - Service-oriented vision |
| Define SOA Adoption Strategy | - Stakeholders<br>- Information about target organizations | - SOA readiness assessment<br>- SOA adoption strategy<br>- Identification of early adopters |
| Develop Service-Oriented Policy Guidance | - Existing policies<br>- Service-oriented vision | - Service-oriented policies<br>- Policy implementation action plan |

*Table 2. Federated Enterprise Architecture processes, Inputs and Outputs*

| Processes | Inputs | Outputs |
|---|---|---|
| Define Service-oriented principle | - Service-oriented vision<br>- Standardized set of SOA features | - Service-oriented principles |
| Identify SOA related standards | - Service-oriented vision | - Identified SOA standards and guide-lines |
| Identify the service-oriented target state | - Service-oriented vision<br>- Asset inventory<br>- Identified SOA standards and guidelines | - Service-oriented target state<br>- Business architecture<br>- Information systems service require-ments<br>- Information systems service architec-ture<br>- SOA integration strategy<br>- Near-term SOA projects<br>- Enterprise Sequencing Plan |

## Federated Enterprise Architecture

Within a SOA-based enterprise, a federated enterprise architecture defines the current and target state of the enterprise. Following the federated model, the enterprise architecture diagrams, views and illustrations allow a mechanism to provide an understanding of individual services, service orchestrations that string along throughout the enterprise, and the enterprise bus infrastructure that allows services to be produced and consumed in an easy manner. Table 2 provides the processes, inputs and outputs for the federated enterprise architecture activities.

## Program Management

This program management activity addresses issues such as how SOA should be governed within the enterprise, what would be a capable governance model, and how do we ensure that the organization follows a system development life cycle that is catering to the dissemination of services. The second aspect

*Table 3. Program Management processes, Inputs and Outputs*

| Processes | Inputs | Outputs |
|---|---|---|
| Develop SOA Governance and Management Strategy | - Service-oriented vision<br>- SOA policies | - SOA governance and management strategy<br>- Updated SOA policies |
| Manage Organizational Change | - Service capability needs analysis<br>- SOA component characteristics<br>- Business capability descriptions<br>- Service-oriented vision<br>- SOA readiness assessment<br>- SOA adoption strategy<br>- Identification of early adopters<br>- Policy implementation action plan<br>- Service-oriented principles<br>- SOA standards, guidelines and conventions<br>- SOA target state of enterprise<br>- SOA integration strategy<br>- SOA governance policies | - Organizational assessment<br>- CM strategic approach<br>- Communication framework<br>- CM/Communications implementation plan |

*Table 4. Portfolio Management processes, Inputs and Outputs*

| Processes | Inputs | Outputs |
|---|---|---|
| Manage Investment Portfolio | - Investment portfolio | - Updated Investment portfolio |

is to figure out how program management activities manage organizational change. This is a very broad topic that covers personnel, facilities and environment, and an approach to configuration management. Table 3 provides the processes, inputs and outputs of the program management activities.

## Executive Portfolio Management

In large enterprises, each organization should have its own portfolio management activities. This allows business and technology stakeholders to define a portfolio of capabilities that the government agency

*Figure 3. Service Enablement Life Cycle Processes*

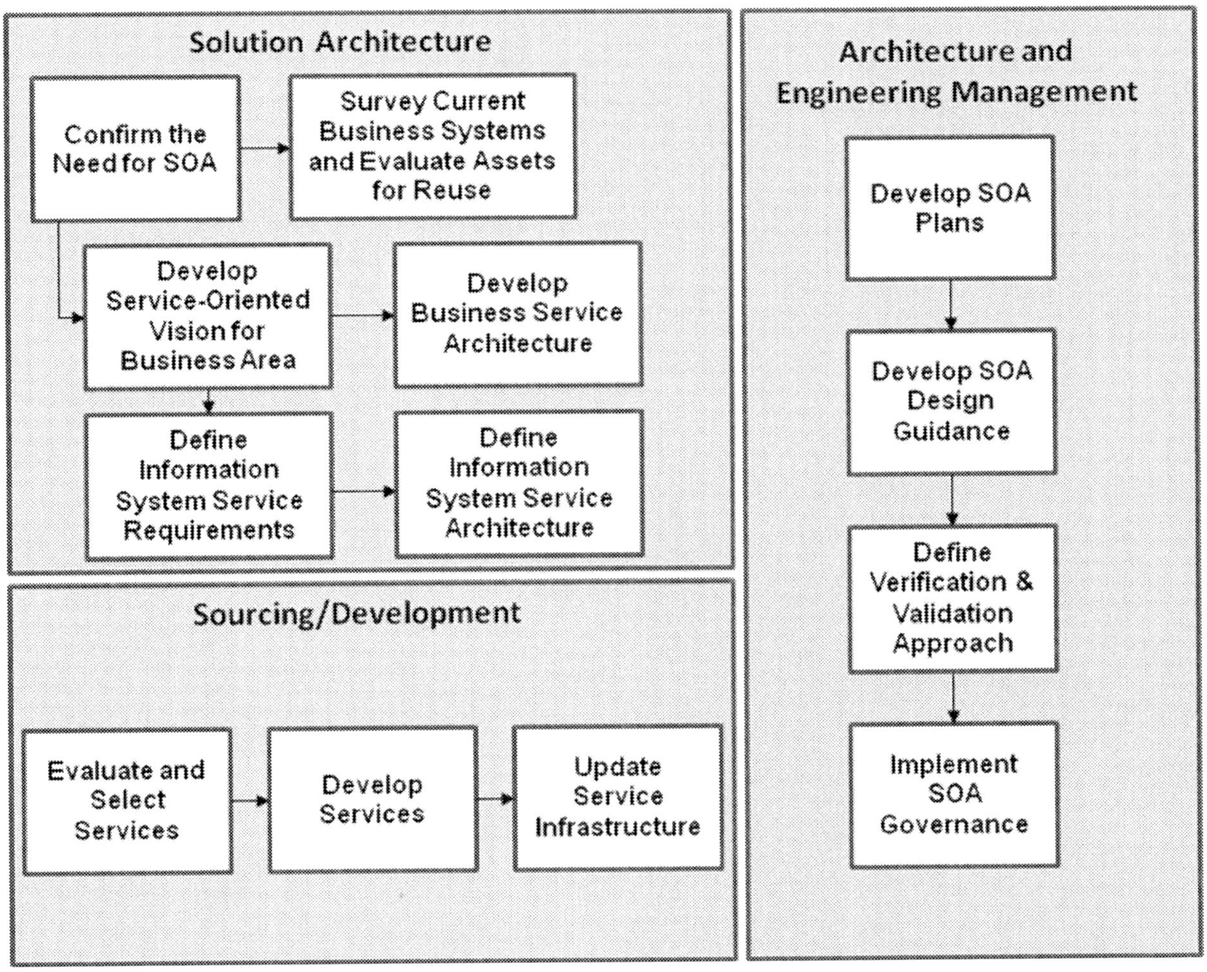

*Table 5. Solution Architecture processes, inputs, outputs*

| Processes | Inputs | Outputs |
|---|---|---|
| Confirm the Need for SOA | - Enterprise-level SOA Readiness Assessment<br>- Enterprise-level SOA Implementation Strategy | - Prepare documented need for SOA |
| Survey Current Business Systems and Evaluate Assets for Reuse | - Asset Inventory | - Documented inventory of reusable business systems and assets |
| Develop Service-Oriented Vision for Business Area | - Service-oriented vision<br>- Service-oriented policies | - Documented service-oriented vision |
| Develop Business Service Architecture | - Information on the service-oriented target state | - Business service architecture |
| Define Information System Service Requirements | - Information system requirements for external organizations | - Service Requirements documentation |
| Define Information System Service Architecture | - Business service architecture | - Information system service architecture |

provides to its citizens. In a SOA-based environment, it is important to identify the portfolio items from the service perspective and review the funding and investment management approaches. Table 4 presents the portfolio management activities.

## Service Enablement Processes

Now that we have discussed the higher level business roadmap, here is an understanding of Service Enablement that provides the process of SOA delivery. Figure 3 provides a set of life cycle processes that allows SOA to be implemented within a large enterprise.

The processes denote three main groups: Solution Architecture, Architecture and Engineering Management, and Sourcing/Development.

## Solution Architecture

Solution architecture allows technology management within the organization to formulate how a Net-Centric SOA environment can be implemented within the enterprise. Here are the processes, inputs and outputs for enabling a service-oriented enterprise that addresses business stakeholder needs and

*Table 6. Sourcing/Development processes, inputs, outputs*

| Processes | Inputs | Outputs |
|---|---|---|
| Evaluate and Select Services | - Asset inventories<br>- Service registries and catalogs<br>- Service requirements and architecture | - Sourcing analysis for reusable services |
| Develop Services | - Business service architecture<br>- Information system service requirements and architecture | - A completed service registry<br>- Service specification |
| Update Service Infrastructure | - Business service architecture<br>- Information system service requirements and architecture | - Logical and Physical Service Infrastructure |

*Table 7. Architecture and Engineering Management processes*

| Processes | Inputs | Outputs |
|---|---|---|
| Develop SOA Plans | - SOA Adoption Strategy<br>- Information system service architecture<br>- Business roadmap sequencing plan | - SOA Adoption Plan<br>- SOA-based Transition Plan and Sequencing Strategy |
| Develop SOA Design Guidance | - SOA principles<br>- Net-centric principles<br>- Information assurance principles | - Service-oriented enterprise design guidance |
| Define Verification and Validation Approach | - Information system service requirements and architecture | - Net-centric SOA verification and validation plan |
| Implement SOA Governance | - Enterprise management and governance strategy<br>- Existing policies | - Governance approach<br>- Policies and Procedures |

prepares appropriate solutions. Table 5 provides the processes, inputs and outputs of the solution architecture.

## Sourcing/Development

This step is when you have to either conduct source selection for hiring contractors to create services for the enterprise, or build each of your services in-house. This includes the ability to evaluate which services should be built, develop those services and then update your service infrastructure to create a service-oriented enterprise. Table 6 provides processes, inputs and outputs of the sourcing and development activities.

## Architecture and Engineering Management

This activity provides you with the technical oversight to conduct your service development activities. Architecture and engineering management tasks include developing SOA plans, developing SOA guidance for software design, defining verification and validation techniques and approaches, and implementing a SOA-based governance. Table 7 provides processes, inputs and outputs of the architecture and engineering management activities.

## Service Sustainment Processes

The next step as part of the transition life cycle is to be able to sustain this new net-centric SOA-based environment. This leads to maintaining tomorrow's enterprise and the ability to evolve further based on the latest technology insertion activities.

Figure 4 provides a model for sustainment of the Net-Centric Service Oriented Enterprise (NCSOE). It provides a set of necessary functions when an organization is engaging in a sustainment activity that implements business and consumer information services.

As we achieve a NCSOE environment, we now have to ensure that we can maintain it properly for our organization. As depicted within Figure 3, this sustainment model includes all of the life cycle processes that were part of the transition mechanism to get us to the target state.

The model accounts for the following set of service sustainment activities:

*Figure 4. Service-Oriented Enterprise Sustainment Model*

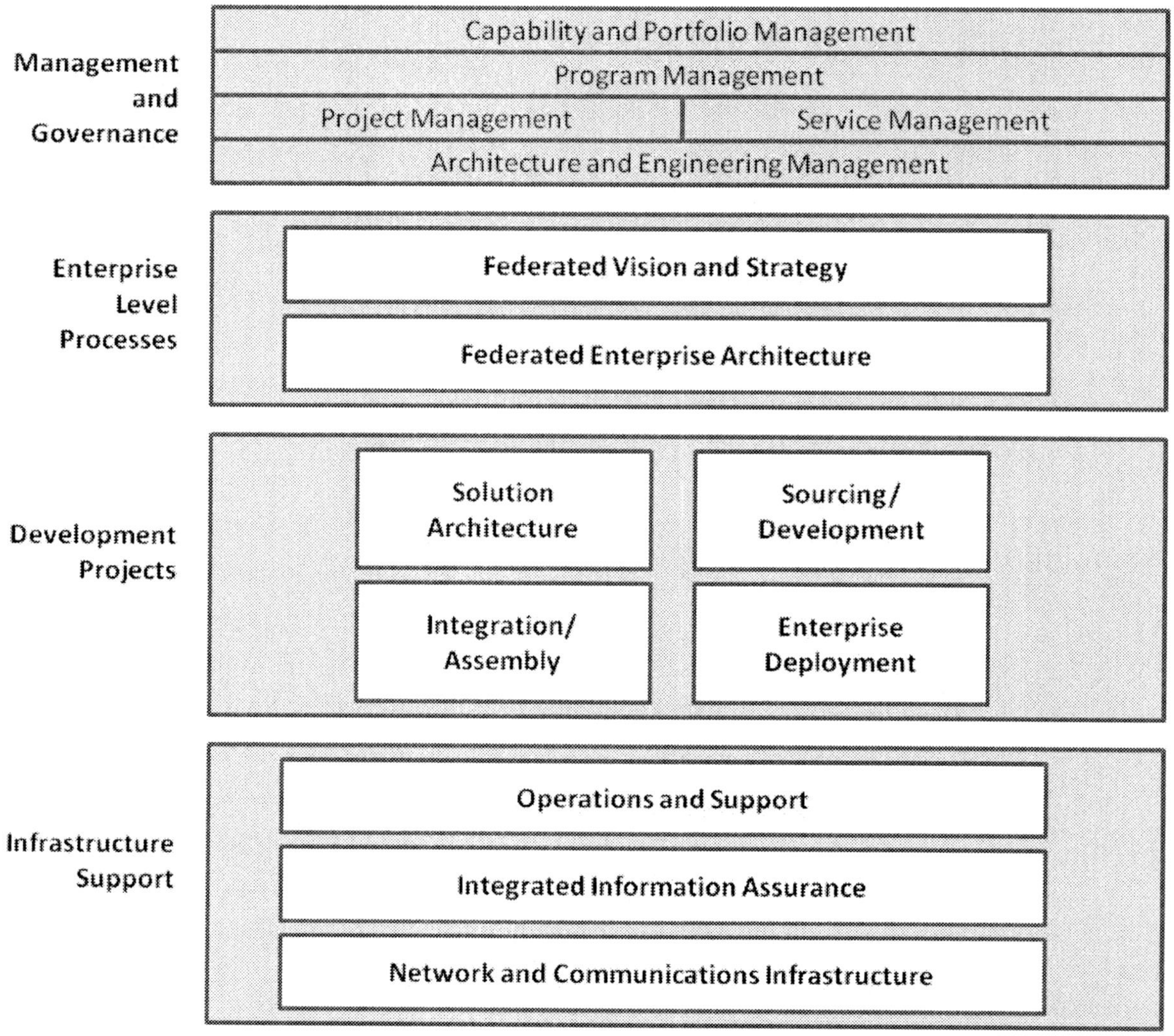


- **Management and Governance Processes** - Enables organizations to maintain their new information enterprise, their ability to quickly meet business and customer needs and their need for ongoing governance.
- **Enterprise Level Processes** – It is useful to ensure that organizations abide by a federated vision and enterprise architecture to account for higher-level and external life cycle processes.
- **Development Projects** - The NCSOE is characterized by a need for rapid and ongoing development that is continually building solutions and services for customers and end-users, and deploying them across an agile enterprise.
- **Infrastructure Support** – This includes a variety of operations and support needs that includes the creation of an integrated information assurance support that eliminates the needs of information silos, and the use of a broadband communications and network infrastructure.

## SERVICE ORIENTED ENTERPRISE TECHNOLOGY FEATURES

As we provide more definition of the NCSOE environment, it is best to provide a set of features that we envision that the NCSOE target state contains. These features are a combination of concepts that we have discussed throughout the rest of the book. However, in this case, these features can be implemented by program management or leadership within government agencies or corporate firms in their upcoming enterprise.

### User Level Agreements

Most enterprises are familiar with service level agreements (SLAs) that exist for most information systems and applications. In this target state, user level agreements provide a twist to SLAs that also incorporate user expectations, usage of services, and the push and pull between service consumers and service producers. A user level agreement would include the end user, the service consumer and the data provider. An end user can access net-centric services via a web-based portal environment. These services that can be available based on collaboration tools, people discovery, content discovery, or a metadata registry. Service consumers can be government agencies or corporate firms that need to conduct business operations. Data providers can be system owners or system developers that support the large amounts of data within the enterprise.

### Service-Oriented System Infrastructure

To properly implement a SOA-based environment, the enterprise is going to require a service-oriented system infrastructure. An infrastructure service can be typically defined as a collection of interface specifications that can be provided by an enterprise resource to other resources and consumers. For an organization, these features now need to be classified as a set of capabilities or a set of system utilities that are being exchanged between providers and consumers. These infrastructure features may include storage and retrieval, configuration, fault-tolerance, performance management, service level agreements, lifecycle resource management.

### Network and Service Monitoring

In enabling a more flexible enterprise, we need to enter a new era when service providers not only tactically respond to problems but also strategically anticipate user needs. In a NCSOE, service providers are no longer monitoring based on a limited set of service parameters. They are able to address enterprise network challenges with a much higher degree of accuracy and completeness for better and faster end-to-end service performance.

### Transport Services

In a NCSOE environment, the current abilities of transport services are enhanced based on features such as mobility, network awareness, situation reactivity, community autonomy, self-learning, service intelligence and management flexibility. Transport services for upcoming IPv6 networks, for example, will require more careful planning and a structured deployment to address these new abilities.

### Service Quality Management

Service quality for a network includes planning and configuring performance such as bandwidth, delay, data loss and traffic priority. The upcoming target state has to have an effective quality of service policies that includes a policy registry and an enforcement function that leverages open standards. Service quality management works in concert with monitoring functions that provide quality indicators. Within a network, these indicators can permit adjustment functions to allow different amounts of traffic to be introduced within the network based on quality of service levels.

### Integrated Information Assurance

The target state has to provide integrated information assurance so that from the user front, there is a common operating picture of the information enterprise for reporting and operational awareness. The information assurance process includes protection, detection, and capability to provide information restoration. The integrated nature allows service consumers and operators to obtain trusted information from across the enterprise acquire in a rapid and assured manner.

### Service Knowledge Management

In a large enterprise, where there is collaboration occurring between a number of different organizations, the collaborated service agreement for an NCSOE environment can become very sophisticated. A dynamic and continuous model that allows the enterprise to detect, monitor and act on an instantaneous basis bridges the knowledge gap. Service knowledge management allows knowledge sharing to occur real-time and provides cost and time benefits. Decision making in a knowledge sharing environment is heavily influenced by dynamic patterns of collaboration and is associated with different levels of accountability.

## EXAMPLE SET OF NET-CENTRIC SERVICES

This section provides an example set of services that is service enabled for a NCSOE. These would undergo the service enablement processes that we discussed earlier for building a set of services for an organization. These service capabilities enable different organizations to collaborate and promote information sharing that leads to greater interoperability.

### Collaboration

This service provides any number of authorized users operating on different computers using web browser to be able to collaborate with each other through voice, video, instant messaging/chat messages, and document sharing.

### Service Discovery (Yellow Pages)

This a searchable repository of services that provides an asset management capability for services within the DoD and supports the full life-cycle of services and service artifacts Provides service providers the capability to publish and advertise service specifications, metadata, and service accessibility to the entire DoD.

### Machine-to-Machine (M2M) Messaging

This allows DoD software applications to interoperate in order to perform synchronous and asynchronous messaging via Web Services. The use of M2M Messaging will provide the DoD with the infrastructure to provide a reliable bridge between multiple organizations for the interoperable sharing of data.

### Mediation

The Mediation service shall provide the means to perform information transformations, protocol adaptation, and orchestration. This will provide the DoD with the infrastructure to translate messages that allows the sharing of data between multiple organizations.

### Enterprise Service Management (ESM)

Enterprise Service Management (ESM) capability shall provide the ability to measure Service Level Agreement (SLA)-based Service Level Metrics, know the situational status of services, and provide service status information to external consumers such as the GIG Infrastructure Services Management Center (GISMC). ESM shall provide the ability to manage and collect NCES service performance information to provide operational support to facilitate root cause analysis of errors and to support government auditing of service performance.

### Service Security

This provides both a Robust Certificate (using public key infrastructure certificates) Validation Service and an Attribute Retrieval Service (using authoritative attributes, e.g., security clearances, etc.) for making security policy decisions to access data, services, and applications.

### Metadata Discovery

This capability discovers and manages (publish, make visible, and access) various metadata artifacts critical to a system and/or person's ability to exchange and understand data components within the enterprise.

### People Discovery

These are equivalent to White Pages that provides for uniquely identifying, finding, and publishing white pages information on people. Provides a web-service interface to the information currently available using the Joint Enterprise Directory Service.

### Content Discovery

This provides a way to perform federated searches for enterprise content across federated search enable data sources. Content Discovery provides the capability to query enterprise content (web pages, text documents, and metadata) which has been exposed to the enterprise by the producer.

### Content Delivery

Content Delivery Service provides users fast access by caching content close to the user and by providing optimized routing of content. The Content Delivery Service provides a smart-cache method of information transport to facilitate the movement of content around the enterprise.

## CHANGES AND BENEFITS TO DIFFERENT INDUSTRY SECTORS

Now that we have defined the future state as a Net-Centric Service Oriented Enterprise, it is important to identify which industry sectors may have an immediate effect. In today's enterprise, most of us would agree that we still can't easily get access to the data and the systems we need in order to get our daily work done. Many of us still spend a great deal of time copying and pasting data between different applications. Critical data is batched, imported and exported between different information systems all across the globe, and it often requires a great deal of manual labor to ensure that the data ends up residing in the correct place. At this time, a number of businesses and corporate firms are pursuing initiatives that implement a service oriented environment. However, most efforts still fallen profoundly short of our desired levels of integration and our expectation for business agility.

### Addressing Business Agility

One of the key challenges that we have today is the problem of inflexibility. As depicted within Figure 29, today's systems have a degree of complexity, they have been designed as information silos, and they are limited in scope since they do not easily allow information to be retrieved and shared. To achieve business agility, large organizations and corporate enterprises need to be able to respond to changes and leverage opportunities for competitive advantage. Figure 29 depicts tomorrow's organization as flexible and empowering. To allow that to happen, current information systems need to be wholly restructured. New systems that are built have to accommodate the needs for greater interoperability and collaboration with multiple external entities.

The rise of a NCSOE environment would offer business agility and implement a SOA-based environment within corporate and government enterprises. NCSOE would allow businesses to conduct enterprise operations. Operational agility provides a requisite level of speed, cost-effectiveness, accuracy and flexibility to promote organizational prosperity. In a service oriented environment, these factors would be addressed based on processes that connect service consumers with service providers. The return on investment for an enterprise can be maximized based on accommodating these factors and multiple perspectives.

## Addressing Benefits to Different Industry Sectors

Based on the discussion of the target state, here is how a NCSOE environment would benefit the following industry sectors in the near future.

### Military Sector

As we have been saying all along, the benefits of a net-centric enterprise is to allow our military to manage its war fighting abilities based on a strong network of well-informed but geographically dispersed set of forces. An NCSOE would provide a high-performance information grid that conducts network-centric warfare. The information ability provides a common operating picture that is applicable to all levels of warfare, contributing to the integration of strategy, operations, and tactics. The use of a NCSOE thereby can increase battle-space awareness and yield increased combat power.

### Financial Services Sector

The financial industry is currently reviewing the potential of a service oriented enterprise. In this new environment, traders can exploit their position of power by using information and information technologies to influence prices and delivery speed in trading objects. NCSOE can be used to perform transaction surveillances for financial companies which institute general operational requirements to avoid fraud by monitoring traders' trading patterns.

### Manufacturing Sector

NCSOE is a powerful information environment that can allow manufacturers to be able to share information in near real-time among all relevant departments. Manufacturers can immediately restructure their inventory information that can simultaneously increase flexibility and responsiveness. A direct benefit from the speed of information sharing is a reduced cost of goods sold, because of lean inventory for parts and finished products. The NCSOE environment can provide a consistent and standardized view of production to avoid risk of creating large inventories that lead to excess finished product, or obsolete/overpriced components. The improved ability to capture the needs of service consumers lead to product attributes that meet consumer's desires, contributing to the design of more-attractive products that in turn trims project-delivery timelines.

### Retail Sector

The retail sector is currently very transaction-intensive. Each large retailer is seeking information superiority by combining business knowledge and intelligence with retailing data to achieve precision retailing. In a NCSOE environment, a sensor can be used to scan, collect, and share product information with suppliers in near real-time, enabling suppliers to optimize and control production and distribution, while appropriately managing their individual supply chains. The result is a significant reduction of distribution costs and high profitability.

### E-Commerce Sector

The key to e-commerce applications is that they offer multi-channel capability that results in end-to-end customer support throughout the buying process, and they use different retail channels to achieve continuity. The NCSOE target state coalesce the integration of front-end customer satisfaction with backend supply chain management, along with intelligent decision making. Customer can visit different vendors to complete a holiday shopping list, and this shopping list would invoke a seamless set of processes to provide order fulfillment and customer requests.

### Transportation

The transportation sector is currently undergoing significant changes. The adoption of NCSOE would allow in-transit traceability so that the enterprise can integrate both sensing and transaction capabilities in supporting information networks. The sensing capabilities can generate near real-time awareness on the precise status and locations of shipments. The Quality of Service would be improved based on a higher level of awareness to provide proactive altering of customer status, identifying the root cause of operational problems, and improving real-time operational performance.

## REVIEW OF CHAPTER GOALS

The goals of this chapter were to address:

- How can we define a Net-Centric Service Oriented Enterprise?
  - Purpose is to achieve a loosely coupled, highly interoperable, distributed system of systems environment.
- How should we formulate the transition mechanism to the target state?
  - Define a set of life cycle-based processes that implement a service-oriented enterprise in your organization.
- How do we develop a business roadmap for implementing services
  - Follow a federated vision and strategy, agree on program management principles, a federated enterprise architecture principles and incorporate portfolio management
- How do we address service enablement and service sustainment?
  - Service enablement provides the process for SOA delivery, and sustainment allows services to be maintained, updated, with the proper infrastructure.
- What are example set of services for a distributed environment?
  - These would be Enterprise SOA Foundation, Collaboration, Content Discovery & Delivery, and a common User Access Portal
- Describe in an easy manner a set of net-centric service capabilities
  - Address the concepts of collaboration, service discovery, machine to machine messaging, mediation, enterprise service management, service security, metadata discovery, people discovery, content discovery and content delivery

## REFERENCES


Alberts, D. S., Garstka, J. J., & Stein, F. P. (1999). *Network Centric Warfare: Developing and Leveraging Information Superiority*, (2nd ed.). Washington, DC: CCRP Publication Series.

Dirner, M., Yuan, E., & Blalock, J. (2007). *Realizing the Army Net-Centric Data Strategy (ANCDS) in a Service Oriented Architecture (SOA)*, Crosstalk, 07. http://www.stsc.hill.af.mil

DoD Information Systems Agency (DISA) (2006). *NCES Draft Pilot Participants Guide*.

DoD Information Systems Agency (DISA) (2008). *Net-Centric Enterprise Services Users Guide, Version 1.2*.

Khoshafian, S., (2006). *Change and Innovation in Service Oriented Enterprises*. BPMInstitute.org

U.S. Army Enterprise Solutions Competency Center (2007). *Service Oriented Architecture Life Cycle Management Model*. Computer Sciences Corporation. Office of the Assistant Secretary of Defense for Networks and Information Integration, Department of Defense Chief Information Officer. (2007). *Department of Defense Net Centric Services Strategy* (Version 1.0).

Valdes, R., (2008, June). *Portals: The Original SOA Framework*. Gartner Internet Platform and Web Services.

# Appendix: Acronyms and Glossary Terms

## ACRONYMS

| | |
|---|---|
| AH | Authentication Header |
| ASD-NII | Assistant Secretary of Defense for Networks and Information Integration |
| BA | Battlespace Awareness |
| BPEL | Business Process Execution Language |
| BRM | Business Reference Model |
| C2 | Command and Control |
| C4 | Command, Control, Communications, and Computers |
| CBA | Capabilities-Based Assessment |
| CC | Common Criteria |
| CD | Compact Disc |
| CDD | Capability Development Document |
| CDR | Critical Design Review |
| CDS | Cross Domain System |
| CERT | Computer Emergency Response Team |
| CES | Core Enterprise Services |
| CIO | Chief Information Officer |
| CJCS | Chairman of the Joint Chiefs of Staff |
| CJTF | Combined Joint Task Force |
| CNA | Computer Network Attack |
| CND | Computer Network Defense |
| CNE | Computer Network Exploitation |
| CNO | Computer Network Operation |
| COA | Course of Action |
| COOP | Continuity Of Operations Plan |
| COI | Community of Interest |
| COMSEC | Communications Security |

| | |
|---|---|
| CONOPS | Concept of Operations |
| CONUS | Continental United States |
| CORBA | Common Object Request Broker Architecture |
| COTS | Commercial-Off-The-Shelf |
| CPD | Capability Production Document |
| CPOE | Computerized Patient Order Entry |
| CRADA | Cooperative Research and Development Agreement |
| CRD | Capabilities Requirements Document |
| DAS | Direct Attached Storage |
| DCOM | Distributed Component Object Model |
| DDMS | DoD Discovery Metadata Standard |
| DECC | DISA Enterprise Computing Center |
| DHCP | Dynamic Host Control Protocol |
| DIACAP | DoD Information Assurance Certification and Accreditation Process |
| DISA | Defense Information Systems Agency |
| DITSCAP | DoD Information Technology Security Certification and Accreditation Process |
| DISR | DoD Information Standards Registry |
| DoD | Department of Defense |
| DOTMLPF | Doctrine, Organization, Training, Materiel, Leadership and Education, Personnel, and Facilities |
| DRM | Data Reference Model |
| EA | Enterprise Architecture |
| EAL | Evaluation Assurance Level |
| EOL | End Of Life |
| ECM | Enterprise Content Management |
| EHR | Electronic Health Record |
| EMP | Electromagnetic Pulse |
| ESP | Encapsulating Security Payload |
| EUCOM | European Command |
| EW | Electronic Warfare |
| Gbs | Giga or Billion bits per second |
| GIG | Global Information Grid |
| GIG-BE | Global Information Grid-Bandwidth Expansion |
| GCCS | Global Combat Communication System |
| GCSS | Global Combat Support System |
| GNO | Global Network Operation |

| | |
|---|---|
| GPS | Global Positioning System |
| HAIPIS | High assurance Internet Protocol Interoperability Specification |
| HHS | Health & Human Services |
| HQ | Headquarters |
| HTTP | Hyper-Text Transfer Protocol |
| IA | Information Assurance |
| IC | Intelligence Community |
| ICD | Initial Capability Document |
| IDE | Integrated Device Electronics |
| IOL | Inter-Operability Laboratory |
| IM | Information Management |
| IP | Internet Protocol |
| IPv4 | Internet Protocol version 4 |
| IPv6 | Internet Protocol version 6 |
| IPT | Integrated Product Team |
| ISP | Internet Service Provider |
| ISR | Intelligence, Surveillance and Reconnaissance |
| IT | Information Technology / Transport |
| JCA | Joint Capabilities Area |
| JCIDS | Joint Capabilities Integration and Development System |
| JEC | Joint Enabling Construct |
| JFC | Joint Functional Concept |
| JIC | Joint Integrating Concept |
| JNO | Joint Net-Centric Operation |
| JOC | Joint Operating Concept |
| JS | Joint Staff |
| JTF | Joint Task Force |
| Kbs | Thousand bits per second |
| KM | Knowledge Management |
| KIP | Key Interface Profile |
| KPP | Key Performance Parameter |
| LAN | Local Area Network |
| LDAP | Lightweight Directory Access Protocol |
| LISI | Levels of Information Systems Interoperability |
| LOC | Line of Communication |
| Mbs | Million bits per second |

| | |
|---|---|
| MHS | Military Health System |
| MPEG | Motion Pictures Expert Group |
| NAS | Network-Attached Storage |
| NAT | Network Address Translation |
| NCE | Net-Centric Environment |
| NCE JFC | Net-Centric Environment Joint Functional Concept |
| NCES | Net-Centric Enterprise Services |
| NCO | Network-Centric Operations |
| NCO CF | Network-Centric Operations Conceptual Framework |
| NCOE | Net-Centric Operational Environment |
| NCW | Network Centric Warfare |
| NDS | National Defense Strategy |
| NGA | National Geo-spatial Intelligence Agency |
| NGO | Non-Governmental Organization |
| NM | Network Management |
| NMS | National Military Strategy |
| NOC | Network Operations Center |
| NOSC | Network Operation and Support Center |
| NCOIC | Net-Centric Operations Industry Consortium |
| NCOW | Net-Centric Operations and Warfare |
| NSS | National Security Strategy |
| OCONUS | Outside the Continental United States |
| OMB | Office of Management & Budget |
| OSI | Open System Interconnection |
| OV | Operational View |
| PDF | Portable Data Format |
| PDR | Preliminary Design Review |
| PKI | Public Key Infrastructure |
| PPTP | Point-to-Point Transport Protocol |
| PRM | Performance Reference Model |
| QoS | Quality of Service |
| RF | Radio Frequency |
| SAN | Storage Area Network |
| SDO | Standards Development Organization |
| SLA | Service Level Agreement |
| SME | Subject-Matter Expert |

| | |
|---|---|
| SMI | Security Management Infrastructure |
| SNMP | Simple Network Management Protocol |
| SOA | Service-Oriented Architecture |
| SOAP | Simple Object Access Protocol |
| SOP | Standard Operating Procedure |
| SRM | Service Reference Model |
| SRR | System Requirements Review |
| SV | System View |
| TBD | To Be Determined |
| TCO | Total Cost of Ownership |
| TCP | Transmission Control Protocol |
| TCS | Tele-Communications Service |
| TMIP-J | Theater Medical Information Program-Joint |
| TRM | Technology Reference Model |
| TV | Technical Standards View |
| UAV | Unmanned Aerial Vehicle |
| UML | Unified Modeling Language |
| UDDI | Universal Description Discovery and Integration |
| VA | Veterans Affairs |
| VHA | Veterans Health Administration |
| VPN | Virtual Private Network |
| WAN | Wide Area Network |
| WSDL | Web Services Description Language |
| XML | Extensible Markup Language |
| XSL | Extensible Stylesheet Language |

## GLOSSARY OF TERMS

| Term | Definition |
|---|---|
| **Access** | To interact with a system entity to manipulate, use, gain knowledge of, and/or obtain a representation of some or all of a system entity's resources |
| **Access Control** | Protection of resources against unauthorized access; a process by which the use of resources is regulated by a security policy and is permitted by only authorized system entities according to that policy |
| **Accessible** | The extent to which users can find and use an information-system resource. |
| **Accurate** | The extent to which a transmission/data stream is error free. |
| **Action** | A structured behavior of limited duration. (JCDRP 7/2004) |

| | |
|---|---|
| **Actionable Knowledge** | Information that enables the decision maker to understand the situation and make use of opportunities for effective action. It is information placed in the context of the situation that includes objectives, constraints, courses of action, uncertainties, and cultural influences. |
| **Activity** | A structured behavior of continuous duration. |
| **Agility** | The ability of an organization to respond quickly to demands or opportunities |
| **Applications** | Software program that performs a specific function directly for a user and can be executed without access to system control, monitoring or administrative privileges. Examples include office automation, electronic mail, web services, and major functional or mission software programs.<br><br>The ability to provide a locally resident software program or group of programs that interfaces directly with Joint Force decision-makers and Communities of Interest, which carries out generalized or mission-specific tasks or processes for which a computer is used, i.e., word processing, spreadsheets, graphics, database management, and communications packages. |
| **Assumption** | A supposition on the current situation or a presupposition on the future course of events, either or both assumed to be true in the absence of positive proof, necessary to enable the commander in the process of planning to complete an estimate of the situation and make a decision on the course of action. |
| **Assured** | Having grounds for confidence that an information technology (IT) product or system meets its certainty or security objectives. |
| **Attribute** | A distinct characteristic inherent in or ascribed to an entity; an entity's attributes are said to describe it |
| **Authentic** | Security measure designed to establish the validity of a transmission, message, or originator, or a means of verifying an individual's authorization to receive specific categories of information. |
| **Authentication** | Security measure designed to establish the validity of a transmission, message, or originator, or a means of verifying an individual's authorization to receive specific categories of information |
| **Authoritative** | Recognized by appropriate governing authorities to be valid or trusted (e.g., the United States [U.S.] Postal Service is the authoritative source for U.S. mailing ZIP codes) |
| **Authorized User** | Any appropriately cleared individual with a requirement to access a DoD information system in order to perform or assist in a lawful and authorized governmental function |
| **Automated Information System (AIS)** | For DoD purposes, an AIS application is the product or deliverable of an acquisition program. An AIS application performs clearly defined functions for which there are readily identifiable security considerations and needs that are addressed as part of the acquisition. An AIS application may be a single software application (e.g., Integrated Consumable Items Support); multiple software applications that are related to a single mission (e.g., payroll or personnel); or a combination of software and hardware performing a specific support function across a range of missions (e.g., Global Command and Control System). AIS applications are deployed to enclaves for operations, and have their operational security needs assumed by the enclave. Note that an AIS application is analogous to a "major application", however, this term is not used in order to avoid confusion with the DoD acquisition category of Major Automated Information System. |
| **Autonomous** | Undertaken or carried on without outside control; existing or capable of existing independently; responding, reacting, or developing independently of the whole |
| **Availability** | Timely, reliable access to data and information services for authorized users (percentage) |
| **Battlespace Knowledge** | Data and information gathered from the battlespace that has been analyzed and integrated through the lens of understanding including the impacts of physical, cultural, social, political, and economic factors on military operations. |
| **Battlespace Awareness** | Situational information resulting from the processing and presentation of time sensitive and perishable data relating to the operational environment, including the status and dispositions of friendly, adversary, and non-aligned actors. |

| | |
|---|---|
| **Business Function** | Something an enterprise does, or needs to do, in order to achieve its objectives |
| **Business Process** | The complete response that a business makes to an event. A business process entails the execution of a sequence of one or more process steps. It has a clearly defined deliverable or outcome. A business process is defined by the business event that triggers the process, the inputs and outputs, all the operational steps required to produce the output, the sequential relationship between the process steps, the business decisions that are part of the event response, and the flow of material and/or information between process steps. |
| **Capability** | The ability to achieve a desired effect under specified standards and conditions through combinations of ways and means to perform a set of tasks. |
| **Care-Of-Address (COA)** | A temporary IP address for a mobile device, the COA enables message delivery when the device is connecting from outside its home network. The care-of address identifies the device's current point of attachment to the Internet and makes it possible to connect from a different location without changing the device's permanent IP address: messages sent to the known permanent address are rerouted to the care-of address while the recipient can be reached there. |
| **Casualty Status** | A term used to classify a casualty for reporting purposes. There are five medical casualty statuses: (1) deceased; (2) very seriously ill or injured; (3) seriously ill or injured; (4) incapacitating illness or injury; and (5) not seriously injured |
| **Certification & Accreditation (C&A)** | The standard DoD approach for identifying information security requirements, providing security solutions, and managing the security of DoD information systems. |
| **Cognitive Domain** | This domain exists in the minds of human beings. This domain is influenced by individual intangibles such as training, experience, public opinion, and situational awareness. Most importantly, the Cognitive Domain is where we make decisions and is directly related to intellectual capabilities and developmental levels. Vital characteristics of this domain are those that affect individual and organizational decision-making, to include attitudes, opinions, beliefs, and values, and understanding. |
| **Collaboration** | Working together in a joint effort for the purpose of achieving a shared understanding, making a decision, or creating a product. |
| **Command and Control (C2)** | The exercise of authority and direction by a properly designated commander over assigned and attached forces in the accomplishment of the mission.<br><br>Command and control functions are performed through an arrangement of personnel, equipment, communications, facilities, and procedures employed by a commander in planning, directing, coordinating, and controlling forces and operations in the accomplishment of the mission. |
| **Community of Interest (COI)** | Communities of Interest (COI) is the inclusive term used to describe collaborative groups of users who must exchange information in pursuit of their shared goals, interests, missions, or business processes and who therefore must have shared vocabulary for the information they exchange. |
| **Complete** | Having all necessary parts, elements, or steps |
| **Computing Environment** | Workstation or server (host) and its operating system, peripherals, and applications |
| **Condition** | A variable of the environment that affects performance of a task |
| **Confidentiality** | Assurance that information is not disclosed to unauthorized persons, processes, or devices |
| **CONOPS** | Concept of Operations - The overall picture and broad flow of tasks within a plan by which a commander maps capabilities to effects, and effects to end state for a specific scenario. |
| **Consistent** | Free from variation or contradiction |
| **Construct** | A concept or theory devised to integrate in an orderly way the diverse data on a phenomenon |
| **Consumer** | An entity (human or machine) that makes use of a service to meet a particular need. Credential: data that is transferred to establish a claimed principal identity |

| | |
|---|---|
| **Control Information** | Information required by the Joint Force to regulate personnel and functions in the execution of command intent. It provides the means to measure, report and correct performance. Tactical control information is required in near-real time to support forces engaged in on-going operations. Operational control information is used to direct assigned forces to accomplish specific missions or tasks that are usually limited by function, time, or location. |
| **Controllable** | The extent to which a network manager has the ability to exercise restraint, direction over, or perform diagnosis to ensure optimal function and security; power or authority to guide, monitor, or manage |
| **Core Enterprise Services (CES)** | A small set of net-centric services, whose use is mandated by the DoD CIO, to provide awareness of, access to and delivery of information on the GIG |
| **Correspondent Node** | A peer node with which a mobile node is communicating. The correspondent node may be either mobile or stationary. |
| **Criterion** | A critical, threshold, or specified value of a measure |
| **Current** | In progress or contemporary |
| **Cyber-Situational Awareness** | Continuously access threats and vulnerabilities to the information domain in order to provide assured information and confidence. Cyber-situational awareness is usually used in conjunction with automated computer defense. |
| **Data** | Information without context |
| **Data Asset** | Data asset refers to any entity that is composed of data. For example, a database is a data asset that comprises data records. In this document, "data asset" means system or application output files, databases, documents, or web pages. "Data asset" also includes services that may be provided to access data from an application. For example, a service that returns individual records from a database would be a data asset. Similarly, a web site that returns data in response to specific queries (e.g., weather.com) would be a data asset. |
| **Decision Support Tools** | Tools intended to help decision makers utilize data and models to identify and solve problems and make decisions |
| **Defense-in-Depth** | The DoD approach for establishing an adequate IA posture in a shared-risk environment that allows for shared mitigation through: the integration of people, technology, and operations; the layering of IA solutions within and among IT assets; and, the selection of IA solutions based on their relative level of robustness. |
| **Deployable** | Effort required to relocate system, or organization to a joint theater arena |
| **Designated Approving Authority (DAA)** | The official with the authority to formally assume responsibility for operating a system at an acceptable level of risk. This term is synonymous with Designated Accrediting Authority and Delegated Accrediting Authority |
| **DMZ (Demilitarized Zone)** | Perimeter network segment that is logically between internal and external networks. Its purpose is to enforce the internal network's IA policy for external information exchange and to provide external, un-trusted sources with restricted access to releasable information while shielding the internal network from outside attacks. A DMZ is also called a "screened subnet." |
| **Distributed** | A structure in which the network resources, such as switching equipment and processors, are dispersed throughout the geographical area being served. Note: Network control may be centralized or distributed |
| **Diverse** | Not dependent on a single element or media |
| **Doctrine** | Fundamental principles by which the military forces guide their actions in support of national objectives. It is authoritative but requires judgment in application |
| **Dynamic** | Reacts appropriately to change in system status |

| | |
|---|---|
| **Dynamic Host Configuration Protocol (DHCP)** | A communications protocol. DHCP automates the assignment of IP addresses in an organization's network. DHCP lets a network administrator supervise and distribute IP addresses from a central point and automatically sends a new IP address when a computer is plugged into a different place in the network. |
| **Domain** | A sphere of activity, concern, or function |
| **Effect** | An outcome (condition, behavior, or degree of freedom) resulting from tasked actions |
| **Employable** | Effort required to commence system operation upon arrival in the Joint Operations Area (JOA) |
| **Enclave** | Collection of computing environments connected by one or more internal networks under the control of a single authority and security policy, including personnel and physical security. Enclaves always assume the highest mission assurance category and security classification of the AIS applications or outsourced IT-based processes they support, and derive their security needs from those systems. They provide standard IA capabilities such as boundary defense, incident detection and response, and key management, and also deliver common applications such as office automation and electronic mail. Enclaves are analogous to general support systems. Enclaves may be specific to an organization or a mission, and the computing environments may be organized by physical proximity or by function independent of location. Examples of enclaves include local area networks and the applications they host, backbone networks, and data processing centers. |
| **End-state** | The set of conditions, behaviors, and freedoms of action that defines<br><br>achievement of the commander's objectives |
| **Enterprise Information Environment Mission Area (EIEMA)** | The Enterprise Information Environment Mission Area (EIEMA) is the DoD portfolio of programs, projects, and systems that deliver the EIE. The EIEMA portfolio enables the functions of the other mission areas, and encompasses all communications, computing, information assurance, and core enterprise service systems, equipment, or software that provide a common information capability or service for enterprise use. |
| **Enterprise Services** | The ability to provide well-defined, enterprise network functions that accept a request and return a response through an interface with a user or another service, such as collaboration, messaging, or information discovery and storage. |
| **Enterprise-wide System Engineering (EWSE)** | This Enterprise-wide System Management Engineering (EWSE) responsibility ensures that upcoming information applications and computing resources are meshed properly with the continuing evolution of the GIG. It is known that providing end-to-end interoperability and consistent performance is essential across the range of business, intelligence, and warfighting functions. The EWSE responsibilities include continuous oversight of the GIG's evolution, developing and maintaining the GIG technical baseline, establishing enterprise-wide capabilities to support decision makers, implementing a program compliance management construct, and overseeing enterprise-wide experiments. |
| **Expeditionary** | Supporting a military operation conducted by an armed force to accomplish a specific objective in a foreign country |
| **Exterior Gateway Protocol (EGP)** | A protocol that distributes routing information to the routers that connect networks. |
| **Federation** | 1. Autonomous organizations operating under a common rule set to a common purpose.<br><br>2. Legally-binding framework to establish and maintain trust among autonomous organizations. |
| **Flexible** | Dynamically meets evolving mission requirements. (Scenario/Condition dependent) |
| **Force Application** | The integrated use of maneuver and engagement to create the effects necessary to achieve assigned mission objectives |
| **Foreign Agent (FA)** | (In Mobile IP,) a router serving as a mobility agent for a mobile node. A foreign agent works in conjunction with another type of mobility agent known as a home agent to support Internet traffic forwarding for a device connecting to the Internet from outside its home network. |

| | |
|---|---|
| **Friction** | The amount of organization effort required to bring a certain set of capabilities to bear in a specified amount of time. |
| **Geo-spatial Information** | The concept for collection, information extraction, storage, dissemination, and exploitation of geodetic, geomagnetic, imagery (both commercial and national source), gravimetric, aeronautical, topographic, hydrographic, littoral, cultural, and spatial data accurately referenced to a precise location on the earth's surface. |
| **Global Information Grid (GIG)** | Global Information Grid (GIG): the globally interconnected, end-to-end set of information capabilities, associated processes, and personnel for collecting, processing, storing, disseminating and managing information on demand to warfighters, policy makers, and support personnel. The GIG includes all owned and leased communications and computing systems and services, software (including applications), data, security services, and other associated services necessary to achieve Information Superiority. It also includes National Security Systems as defined in section 11103 of title 40 of the United States Code. The GIG supports all Department of Defense, National Security, and related Intelligence Community missions and functions (strategic, operational, tactical, and business) in war and in peace. The GIG provides capabilities from all operating locations (bases, posts, camps, stations, facilities, mobile platforms, and deployed sites). The GIG provides interfaces to coalition, allied, and non-DoD users and systems. The GIG includes any system, equipment, software, or service that meets one or more of the following criteria: 1) transmits information to, receives information from, routes information among, or interchanges information among other equipment, software, and services; 2) provides retention, organization, visualization, information assurance, or disposition of data, information, and/or knowledge received from or transmitted to other equipment, software, and services; and 3) processes data or information for use by other equipment, software, or services. |
| **Governance** | The systems, processes, and procedures put in place to steer the direction, management, and accountability of an organization. In the context of the SOA in the DoD, governance means establishing and enforcing how DoD Components agree to provide, use, and operate services. |
| **Identity** | The collective set of attributes that defines an entity (i.e., subject, resource, etc.) within a given context. |
| **Information** | Any communication or representation of knowledge such as facts, data, or opinion in any medium or form, including textual, numerical, graphic, cartographic, narrative, or audiovisual forms. |
| **Information Assurance (IA)** | The ability to provide the measures that protect and defend information and information systems by ensuring their availability, integrity, authentication, confidentiality, and non-repudiation. This includes providing for the restoration of information systems by incorporating protection, detection, and reaction capabilities. |
| **IA Control** | An objective IA condition of integrity, availability or confidentiality achieved through the application of specific safeguards or through the regulation of specific activities that is expressed in a specified format, i.e., a control number, a control name, control text, and a control class. Specific management, personnel, operational, and technical controls are applied to each DoD information system to achieve an appropriate level of integrity, availability, and confidentiality. |
| **Information Domain** | Where information exists. The Information Domain has a dual nature, consisting of the information itself and the medium by which we collect, process, and disseminate the information. Characteristics of the Information Domain include information quality (completeness, accuracy, timeliness, relevance, and consistency), distribution (range, sharing, and continuity), and interaction (exchange or flow of information). |
| **Information Management (IM)** | The planning, budgeting, manipulating, and controlling of information throughout its life cycle. |
| **Information Mobility** | The dynamic availability of information. Information mobility is aided or impeded by culture, policy, governance, economics and resources and technology and infrastructure |
| **Information Sharing** | Making information available to participants (people, processes, or systems). Information sharing includes the cultural, managerial, and technical behaviors by which one participant leverages information held or created by another participant. |

| | |
|---|---|
| **Information Transport** | The ability to provide the physical communications media over which assured connectivity takes place, supported by switching and routing systems. |
| **Infrastructure** | All building and permanent installations necessary for the support, redeployment, and military forces operations (i.e., barracks, headquarters, airfields, communications, facilities, stores, port installations, and maintenance stations). |
| **Integrated** | All functions and capabilities focused toward a unified purpose |
| **Integrity** | Quality of an information system reflecting the logical correctness and reliability of the operating system; the logical completeness of the hardware and software implementing the protection mechanisms; and the consistency of the data structures and occurrence of the stored data. Note that, in a formal security mode, integrity is interpreted more narrowly to mean protection against unauthorized modification or destruction of information. |
| **Interdependence** | A mode of operations based on a high degree of mutual trust, where diverse members make unique contributions toward common objectives and may rely on each other for certain essential capabilities rather than duplicating them organically |
| **Intermediate System to Intermediate System (IS-ISv6)** | An OSI/IP routing protocol, IS-ISv6 is the new version that supports IPv6 addressing. MPLS traffic engineering parameters can be distributed with IS-IS using extensions to the protocol (IS-IS-TE). |
| **Internet Control Message Protocol v6 (ICMPv6)** | An extension to IP that allows for the generation of error messages, test packets, and informational messages related to IP. |
| **Internet Gateway Protocol (IGP)** | Protocol that distributes routing information to the routers within a network. The term gateway is historical; router is currently the preferred term. Example IGPs are OSPF, IS-IS and RIP. |
| **Interoperability** | An operational spectrum of compatibility and connectedness that ranges from isolation to integration |
| **Joint** | Connotes activities, operations, organizations, etc., in which elements of two or more Military Departments participate with interagency and multinational partners. |
| **Joint Net-Centric Operation (JNO)** | The ability to exploit all human and technical elements of the Joint Force and its mission partners by fully integrating collected information, awareness, knowledge, experience, and decision-making, enabled by secure access and distribution, to achieve a high level of agility and effectiveness in a dispersed, decentralized, dynamic and/or uncertain operational environment. |
| **Key Performance Parameter** | Those minimum attributes or characteristics considered most essential for an effective military capability. |
| **Knowledge** | Data and information that have been analyzed to provide meaning and value. Knowledge is various pieces of the processed data and information that have been integrated through the lens of understanding to begin building a picture of the situation. |
| **Knowledge Management (KM)** | The systematic process of discovering, selecting, organizing, distilling, sharing, developing and using information in a social domain context to improve organizational effectiveness |
| **Knowledge Sharing** | The ability of networked users to manage and make available relevant, accurate information, transform it into knowledge, and act upon it with confidence. This provides access to newly discovered or recurring information in a usable format and facilitates collaboration, distributed decision-making, adaptive organizations, and a greater unity of effort via synchronization and integration of force elements to the lowest levels. |

| | |
|---|---|
| **Level I Care** | Level I care consists of care rendered at the unit level at the point of injury. It includes self-aid, buddy aid, and combat lifesaver skills, examination, and emergency lifesaving measures such as the maintenance of the airway, control of bleeding, prevention and control of shock, splinting or immobilizing fractures, and the prevention of further injury. |
| **Level II Care** | At a minimum, Level II care includes physician-directed resuscitation and stabilization and may include advanced trauma management, emergency medical procedures, and forward resuscitative surgery. Supporting capabilities include basic laboratory, limited x-ray, pharmacy, and temporary holding facilities. Patients are treated and returned to duty, or are stabilized for movement to a MTF capable of providing a higher level of care. |
| **Level III Care** | Care is administered that requires clinical capabilities normally found in a facility that is typically located in a reduced-level enemy threat environment. The facility is staffed and equipped to provide resuscitation, initial wound surgery, and postoperative treatment. This level of care may be the first step toward restoration of functional health, as compared to procedures that stabilize a condition to prolong life. |
| **Level IV Care** | In addition to providing surgical capabilities found at Level III, this level also provides rehabilitative and recovery therapy for those who can return to duty within the theater patient movement policy. This level of care may only be available in mature theaters. |
| **Level V Care** | Level V definitive care includes the full range of acute convalescent, restorative, and rehabilitative care and is normally provided in CONUS by military and VA hospitals, or civilian hospitals that have committed beds for casualty treatment as part of the National Defense Medical System. |
| **Manageable** | Capable of being controlled, handled, or used with ease |
| **Materiel** | All items (including ships, tanks, self-propelled weapons, aircraft, etc., and related spares, repair parts, and support equipment, but excluding real property, installations, and utilities) necessary to equip, operate, maintain, and support military activities without distinction as to its application for administrative or combat purposes |
| **Measure** | Quantitative or qualitative basis for describing the quality of task performance |
| **Measures of Performance** | Measures designed to quantify the degree of perfection in accomplishing functions or tasks |
| **Measures of Effectiveness** | Measures designed to correspond to accomplishment of mission objectives and achievement of desired effects |
| **Medical Surveillance** | The ongoing, systematic collection of health data essential to the evaluation, planning, and implementation of public health practice, closely integrated with timely dissemination of data required by higher authority |
| **Metadata** | Metadata is descriptive information about the meaning of other data. Metadata can be provided in many forms, including XML. |
| **Metadata Catalog** | Metadata Catalog is a system that contains the instances of metadata associated with individual data assets. Typically, a metadata catalog is a software application that uses a database to store and search records that describe such items as documents, images, and videos. Search portals and applications can use metadata catalogs to locate the data assets that are relevant to their queries. |
| **Metadata Registry** | Metadata Registry is a system that contains information that describes the structure, format, and definitions of data. Typically, a registry is a software application that uses a database to store and search data, document formats, definitions of data. and relationships among data. System developers and applications are the predominant users of a metadata registry. |
| **Metric** | A quantitative measure associated with an attribute |
| **Mission** | The end state, purpose, and associated tasks assigned to a single commander |

| | |
|---|---|
| **Mission Area** | A defined area of responsibility with functions and processes that contribute to mission accomplishment. In the context of managing the DoD's portfolios of GIG investments, the DoD has four major categories of mission areas - the Warfighter Mission Area, the Business Mission Area, the Defense Intelligence Mission Area, and the Enterprise Information Environment Mission Area (EIEMA). |
| **Mission Assurance Category (MAC)** | Applicable to DoD information systems, the mission assurance category reflects the importance of information relative to<br><br>the achievement of DoD goals and objectives, particularly the warfighters' combat mission. Mission assurance categories are primarily used to determine the requirements for availability and integrity. The Department of Defense has three defined mission<br><br>assurance categories:<br><br>Mission Assurance Category I (MAC I) - Systems handling information that is determined to be vital to the operational readiness or mission effectiveness of deployed and contingency forces in terms of both content and timeliness. The consequences of loss of integrity or availability of a MAC I system are unacceptable and could include the immediate and sustained loss of mission effectiveness. MAC I systems require the most stringent protection measures.<br><br>Mission Assurance Category II (MAC II) - Systems handling information that is important to the support of deployed and contingency forces. The consequences of loss of integrity are unacceptable. Loss of availability is difficult to deal with and can only be tolerated for a short time. The consequences could include delay or degradation in providing important support services or commodities that may seriously impact mission effectiveness or operational readiness. MAC II systems require additional safeguards beyond best practices to ensure adequate assurance.<br><br>Mission Assurance Category III (MAC III) - Systems handling information that is necessary for the conduct of day-to-day business, but does not materially affect support to deployed or contingency forces in the short-term. The consequences of loss of integrity or availability can be tolerated or overcome without significant impacts on mission effectiveness or operational readiness. The consequences could include the delay or degradation of services or commodities enabling routine activities. MAC III systems require protective measures, techniques or procedures generally commensurate with commercial best practices. |
| **Mobile Node** | A node that can change its point of attachment from one link to another, while still being reachable via its home address. |
| **Multi-Protocol Border Gateway Protocol Plus (MBGP+)** | MBGP+ enhances BGP to support more types of advertised routes, including IPv6 routes. |
| **Need to Know** | Necessity for access to, or knowledge or possession of, specific official DoD information required to carry out official duties |
| **Net Centricity** | Net-centricity is the realization of a networked environment (including infrastructure, systems, processes, and people) that enables a completely different approach to war fighting and business operations. |
| **Net-Centric Environment (NCE)** | A Joint Force framework for full human and technical connectivity that allows all DoD users and mission partners to share the information they need, when they need it, in a form they can understand and act on with confidence; and protects information from those who should not have it. |
| **Net-Centric Operations (NCO)** | The exploitation of the human and technical networking of all elements of an appropriately trained Joint Force by fully integrating collective capabilities, awareness, knowledge, experience, and superior decision-making to achieve a high level of agility and effectiveness in dispersed, decentralized, dynamic and uncertain operational environments |

| | |
|---|---|
| **Net-Centric Operational Environment (NCOE)** | The coherent application of seamless, integrated net-centric capabilities to the forward edge of the battlespace enabling full spectrum dominance. |
| **Net-Centric Warfare (NCW)** | An information superiority-oriented concept of operations that generates increased combat power by networking sensors, decision-makers, and shooters to achieve shared awareness, increased speed of command, higher tempo of operations, greater lethality, increased survivability, and a degree of self-synchronization. This is a sub-set of Net-Centric Operations. |
| **Net-Centric Environment (NCE)** | The Net-Centric Environment is a framework for full human and technical connectivity and interoperability that allows all DoD users and mission partners to share the information they need, when they need it, in a form they can understand and act on with confidence; and protects information from those who should not have it. (Net-Centric Environment Joint Functional Concept, Version 1.0, April 7, 2005) |
| **Net-Centric Operations and Warfare Reference Model (NCOW-RM)** | The NCOW RM is a reference model that provides a description of enterprise level activities, services, technologies, and concepts (e.g. data strategy) that enable a net-centric environment for cross-banding, business, and management operations. |
| **NetOps** | Provides assured and timely net-centric services across strategic, operational and tactical boundaries in support of DoD's full spectrum of warfighting, intelligence and business missions. NetOps provides an integrated approach to accomplishing three interdependent tasks necessary to operate the GIG: GIG Enterprise Management (GEM), GIG Network Defense (GND), and Information Dissemination Management/Content Staging (IDM/CS). |
| **Network** | Two or more computers connected to each other. The purpose of a network is to enable the sharing of files and information between multiple systems. The Internet can be described as a global network of networks. Computer networks can be connected through cables (Ethernet cables or phone lines) or wirelessly, using wireless networking cards that send and receive data through the air. |
| **Network Management (NM)** | Provides the network with the desired level of quality, agility, and trustworthiness. NM focuses on the configuration, availability, performance and manageability of network services and the underlying physical assets that provide end-user services, as well as connectivity to enterprise application services. |
| **Network Address Translation (NAT)** | The translation of an Internet Protocol address used within one network to a different IP address known within another network. This allows duplicate IP addresses to be used within an organization and unique addresses outside. |
| **Non-Repudiation** | Assurance the sender of data is provided with proof of delivery and the recipient is provided with proof of the sender's identity, so neither can later deny having processed the data |
| **Objective** | A desired end derived from guidance |
| **Ontology** | Ontology includes data categorization schemes, words within thesaurus, vocabularies, key-word lists, and taxonomies. Ontologies promote semantic and syntactic understanding of data. |
| **Open Shortest Path First (OSPF)v3** | A link-state routing protocol used by IP routers located within a single Autonomous System (AS) to determine routing paths. OSPFv3 is the version supports IPv6 addresses. |
| **Operational Readiness** | The capability of a unit/formation, ship, weapon system, or equipment to perform the mission or functions for which it is organized or designed. May be used in a general sense or to express a level or degree of readiness. |
| **Patient Movement** | The act or process of moving a sick, injured, wounded, or other person to obtain medical and/or dental care or treatment. Functions include medical regulating, patient evacuation, and en route medical care. |

| | |
|---|---|
| **Portfolio** | The aggregate of IT investments for DoD information systems,<br><br>infrastructure and related technical activities that are linked to mission goals, strategies, and architectures, using various assessment and analysis tools to permit information and IT decisions to be based on their contribution to the effectiveness and efficiency of military missions and supporting business functions. Portfolios enable the Department of Defense to manage IT resources and align strategies and programs with Defense-wide, functional, and organizational goals and measures. |
| **Proxy** | Software agent that performs a function or operation on behalf of another application or system while hiding the details involved. Typical proxies accept a connection from a user, make a decision as to whether or not the user or client network address is authorized to use the requested service, optionally perform additional authentication, and then complete a connection on behalf of the user to a remote destination. |
| **Public Domain Software** | Software not protected by copyright laws of any nation that carries no warranties or liabilities, and may be freely used without permission of or payment to the creator. |
| **Quality** | Lacking nothing essential or normal |
| **Risk** | Probability and severity of loss linked to hazards |
| **Risk Management** | The process of identifying, assessing, and controlling risks arising from operational factors and making decisions that balance risk cost with mission benefits |
| **Robust** | Having or exhibiting strength or vigorous health |
| **Routing Information Protocol next generation (RIPng)** | An Internet routing protocol that uses hop count as a routing metric. RIP is the most common IGP used in the Internet. RIPng is the new version that supports IPv6 addresses. |
| **Schema** | Schema is a diagrammatic representation, an outline, or a model. In relation to data management, a schema can represent any generic model or structure that deals with the organization, format, structure, or relationship of data. Some examples of schemas are:<br><br>(1) a database table and relationship structure,<br><br>(2) a document type definition (DTD),<br><br>(3) a data structure used to pass information between systems, and<br><br>(4) an XML schema document (XSD) that represents a data structure and related information encoded as XML. Schemas typically do not contain information specific to a particular instance of data. |
| **Service** | A mechanism to enable access to one or more capabilities, where the access is provided using a prescribed interface and is exercised consistent with constraints and policies as specified by the service description. |
| **Service Oriented Architecture** | A paradigm for defining, organizing, and utilizing distributed capabilities in the form of loosely coupled software services that may be under the control of different ownership domains. It provides a uniform means to offer, discover, interact with, and use capabilities to produce desired effects that are consistent with measurable preconditions and expectations. |
| **Service Provider** | An entity (i.e., person or organization) that offers the use of capabilities by means of a service. |
| **Shared Understanding** | A shared appreciation of the situation supported by common information to enable rapid collaborative joint engagement, maneuver, and support |
| **Shared Space** | Shared space is a mechanism that provides storage of and access to data for users within a bounded network space. Enterprise-shared space refers to a store of data that is accessible by all users within or across security domains on the GIG. A shared space provides virtual or physical access to any number of data assets (e.g., catalogs, web sites, registries, document storage, and databases). As described in this Strategy, any user, system, or application that posts data uses shared space. |

| | |
|---|---|
| **Situational Awareness (SA) Information** | An information category depicting individual Joint Force members' perceptions of the current environment. Based upon gathered, accessible, and provided data, it allows individuals to identify where friendly forces are in relation to each other. Additionally, SA includes the location of enemy and neutral forces within the battlespace and what those forces are doing. The SA information timing is near instantaneous, allowing the Joint Force the ability to make rapid and decisive decisions based upon that information. |
| **Social Domain** | The Social Domain is shaped by the specifics of language and symbolic communication among human beings. It is the domain within which individuals interact and is strongly influenced by tacit knowledge including elements of culture, education, collective experience, and morale. |
| **Stakeholder** | A DoD entity with a direct interest, involvement, and investment in DoD information sharing |
| **Standard** | The minimum proficiency required in the performance of a task. For mission essential tasks of Joint Forces, each task standard is defined by the Joint Force commander and consists of a measure and criterion. |
| **Synchronization** | The arrangement of military actions in time, space, and purpose to produce maximum relative combat power at a decisive place and time and in the intelligence context, application of intelligence sources and methods in concert with the operation plan. |
| **System** | A regularly interacting group of items forming a unified whole |
| **Task** | An action or activity defined within doctrine, standard procedures, or concepts that may be assigned to an individual or organization |
| **Threshold (value)** | A minimum acceptable operational value below which the utility of the system becomes questionable |
| **Trustworthy** | The extent to which confidence or assurance is held in information or decisions |
| **Understanding** | Knowledge that has been synthesized and had judgments applied to it in the context of a specific situation. Understanding reveals the relationships among the critical factors in any situation |
| **Vignette** | A concise narrative description that illustrates and summarizes pertinent circumstances and events from a scenario |
| **Web Service** | Web services are self-describing, self-contained, modular units of software application logic that provide defined business functionality. Web services are consumable software services that typically include some combination of business logic and data. Web services can be aggregated to establish a larger workflow or business transaction.<br><br>Inherently, the architectural components of web services support messaging, service descriptions, registries, and loosely coupled interoperability. |
| **Workflow** | A graphic representation of the flow of work in a process and its related sub-processes; including specific activities, information dependencies, and the sequence of decisions and activities. |
| **XML** | Extensible Markup Language (XML) is a tagging language used to describe and annotate data so it can be consumed by human and system interactions. XML is typically arranged hierarchically using XML elements and attributes. It also uses semantically rich labels to<br><br>describes elements and attributes to enable meaningful comprehension.<br><br>An example of XML data describing an element named "Person" appears as follows:<br><br><Person><br><br><FirstName>John</FirstName><br><br><MiddleInitial>H</MiddleInitial><br><br><LastName>Doe</LastName><br><br></Person> |

# About the Authors

**Supriya Ghosh** has always perceived life from a global perspective. He was born in Calcutta, India, spent his early childhood in London, England and then grew up in New York City. He attended and received his BS from California Institute of Technology, in Pasadena, CA and then an MS in aeronautics and astronautics from the University of Washington, in Seattle, Washington. He spent his early career in the aerospace industry in Los Angeles as a propulsion engineer building rockets. He then went on to found a multimedia consulting firm, helping Hollywood enter the digital information industry. Mr. Ghosh then moved to the Washington DC area working as a technology consultant to the federal government, along with the healthcare and financial sectors. He received an MBA from Johns Hopkins University. Over the years, Ghosh has become an expert in the area of enterprise architecture. He served as principal prchitect for Lockheed Martin and was responsible for providing strategic and technical direction for federal programs. He founded his own technology consulting firm, Arcadia Concepts LLC helping the Department of Defense. In 2008, Ghosh was lead architect for the Presidential mandate affecting Wounded Warriors to develop a joint DoD/VA Digital Image Sharing Plan for essential healthcare images. Ghosh is currently serving as enterprise architecture Manager for the U.S. Army Acquisition domain, and providing thought leadership to a number of organizations.

**Dan Adams** is a military subject matter expert and works for the U.S. Army conducting acquisition enterprise requirements, plans and programs, statutory and regulatory financial management issues. Adams retired as Lieutenant Colonel from the U.S. Army. He is a graduate of the University of Arkansas at Little Rock, where he received his commission as a distinguished military graduate. He is an experienced military parachutist, and a qualified parachute rigger. He was commissioned as a regular army officer, and after attending initial and specialized training, was assigned to Berlin Brigade serving in the 2d Battalion, 6th Infantry. From 1987 to 1991, he was assigned to Fort Bragg serving in units of 18th Airborne Corps to include the 82d Airborne Division and the Corp Staff. Lieutenant Colonel Adams gained key operational experience as a Plans Officer Joint Task Force–Bravo in Honduras, and as the G-4 JTF-140 on Saint Croix following Hurricane Hugo and as a company commander in Saudi Arabia and Iraq during Desert Shield/Desert Storm. While assigned to Fort Bragg he spent significant time in Central America, and South West Asia. He has held key leadership positions and has worked for U.S. Army headquarters as a policy maker for a number of years.

**Russell S. Boyd** serves as the practice lead for enterprise architecture and investment in the defense marketplace with an emphasis on enabling government agencies to achieve IT legacy modernization, business transformation and information sharing objectives. Boyd has extensive leadership and management experience with information technology, interfaces, system integration, large system implementations and upgrades. Boyd has been practicing enterprise architecture and has experience in codifying and proliferating technical standards for the DoD enterprise. Boyd is currently involved in preparing net-centric enterprise architecture views for organizations. Boyd is a retired Air Force officer who was active in the Air Force Medical Corps. His military career included managing the day-to-day information technology operations for DoD Military Health System hospitals and clinical laboratories.

**Mark G. Janczewski** is an expert in military medicine and health information technology. He is a board-certified physician and has an extensive medical and educational background having received numerous academic degrees. He is a retired Colonel from the U.S. Air Force and currently consults for the DoD and Department of Veterans Affairs. His career of over 35 years spans a wide range starting with computer communications working for the MITRE Corporation and as a US Army officer in the Signal Corps. Dr. Janczewski entered the Air Force as a primary care physician at Rhein-Main AB near Frankfurt. He became more involved in medical information systems and tele-health working for the Air Force Surgeon General as the first chief of the AFMS Advanced Informatics Strategy Office and director of Air Force Telemedicine. He was director of development for MHS Clinical Information Technology Program Office, the first chief of DoD Telehealth and deputy chief of e-Health requirements and participated in rolling out the military's electronic health record. Dr. Janczewski received a BS in mathematics from Virginia Polytechnic Institute, an MA in math from Georgetown University, an MD from Eastern Virginia Medical School, and an MPH from Johns Hopkins University. Dr. Janczewski is president of Medical Networks LLC, his own consulting firm providing clinical information technology subject matter expertise to government and corporate clients.

# Index

## E



## F



## G

## H



## I

## J



## K



## L



## M



## N

## O



## P



## Q

## R



## S



## T



## U

## V



## W



## X